# PRINCIPIA MATHEMATICA

## An introduction to the Absolute Geometry of Space-Time and Matter

**Besud Ch. Erdeni**

I consider natural philosophy rather than arts and write not concerning manual but natural powers...
**Sir Isaac Newton**

authorHOUSE®

*AuthorHouse™ UK Ltd.*
*1663 Liberty Drive*
*Bloomington, IN 47403 USA*
*www.authorhouse.co.uk*
*Phone: 0800.197.4150*



*Published by AuthorHouse*

*ISBN: 978-1-4817-9806-8 (sc)*
*ISBN: 978-1-4817-9807-5 (e)*

*This book is printed on acid-free paper.*

## What is Logic ?

Even logical consequences just happen.
**Henry S. Haskins**

Let us reformulate Newton's First Rule of Reasoning in an extreme miminax form as follows:

**Logic is what gets the greatest possible by the least possible.**

The least is the mathematical point, that is, **Nothing**, while the greatest is the Universe, that is, **Something**.

Consequently, a single point has to be both nesessary and sufficient to reconstruct the Cosmos on paper.

## What is Space and What is Time ?

To describe right lines and circles are problems, but not geometrical problems. The solution of these problems is required from mechanics.
**Sir Isaac Newton**

In pure theories is given the only initial condition, notably, a point resting on a coordinate plane. Its translational motion (bifurcation) will draw a line segment . Denote it by the Golden section constant

(1) $$\Phi = \frac{1+\sqrt{5}}{2} = 1.618\ 033\ 988\ 749...$$

The Golden ratio was known as early as to the ancient Sumerians, let alone the antique Greeks. Despite this, the following is a discovery

(2) $$X^{\Phi \pi e} X = 10^{90},$$

where is seen, so to say, the Hole Trinity of the mathematical continuum, or else, uncountably infinite set of real numbers. Thus, we derive a new fundamental constant

(3) $$X = 1185403.539676801580...$$

Intuitively, it should display some harmonizing properties with respect to whole numbers. Indeed,

(4) $$X = \frac{\mathbf{162400285}}{\mathbf{137}.0000000...};$$

(5) $$\frac{X}{\mathbf{163}} \cdot \frac{360}{\Phi^2} = 10^{6.00000...}.$$

Now that the Φ line rotates around the origin and bifurcates into

(6) $$i_1 = \sqrt{\Phi},$$

as shown below (Picture A):

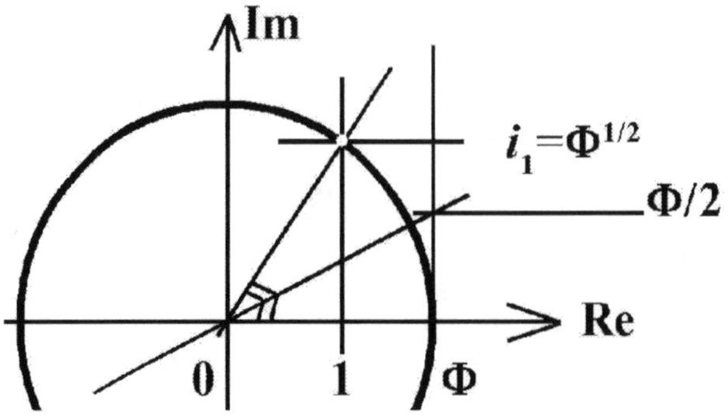


As a result of the previous two sequential bifurcations, we may postulate that

(7) $$\Phi - \textit{Newton's absolute mathematical space};$$
$$i_1 - \textit{Newton's absolute mathematical time}.$$

Then, the four-dimensional space-time shall be written in the form

(8) $$\Phi^3 i_1 = 5{,}3883617040...$$

All the mysteries of space and particularly that of time dimension are going to be replaced with the previous entity which is merely a number, however. According to Newton's own view, space and time do exist and as such they require for some absolute notations. To denote space and time absolutely, we would need in numbers of unique properties. This unique number is notably the constant, Φ.

Anyway, we have began to construct geometry from the beginning and, if logically, this new geometry has to have no more than just one postulate. The latter reads: **The real geometry of space-time is Φ-invariant**.

The entity (8) is the space-time point implying that geometry becomes quantized, that is, discretized. Consequently, we will deal with the space-time discontinuum. The cruciality of discretization of geometry has been first pointed out by Henry Poincare a century ago. And this was the key problem of sciences thoroughout the 20th century.

What requires for discernment is, notably, the minimax logic (MML) we have formulated at the beginning. According to it, this geometry has only one figure: mathematical point. Consequently, the Universe was and still remains the same point-singularity made with a wealth of properties we will investigate in this work.

Besides, Einstein's view on geometry does not, in fact, contradict to Newton's, for the former spoke of the importance of deriving the space and time scales from more fundamental principles than, say, special relativity theory.

The moral lesson of absolute geometry is that the classics of the exact sciences should be read very carefully; otherwise, the process of developing fundamental theories risks to stagnate under the cover of false sensations, as it did actually happen during the later decades of the past century. The Euclidean geometry is a spatial geometry. How it can be completed with temporal dimension? This question came to the author's mind 50 years ago. As it is known, any good discovery is the answer to a foolish question.

## A Proof of the New Geometry

Die ganzen zahlen hat der liebe Gott gemacht,
alles andere est Menschenwerk.

**Leopold Kronecker**

The existing quantum-relativistic theories a litte bit imprecise compared with the experimental data. This situation reveals itself in the anomaly of the electron magnetic moment which does notably define what we know in physics and what do not know yet. Reference sources provide the experimental value of the magnetic moment anomaly as being close to

(9) $$a_{Electron} = 0.001159\,652\,1$$

Right at the moment we have learned what is exactly space-time in Newton's or any other terms. Therefore, immediately,

(10) $$\frac{1}{a_{Electron}} = \sqrt[\Phi^3 i_1]{\mathbf{658361} \cdot 10^{10}} \, .$$

Whence pure-theoretically

(11) $$a_e = 0.0011596520997696346686715329576713 \, .$$

It is going to be true if only the integer, **658461**, is not by a blind chance. If it arises reasonably, then it should, intuitively, be a most fundamental number with particular properties. Indeed,

(12) $$e^{5\Phi \pi i_1 e\sqrt{2}} \cdot \mathbf{658361} = \frac{\mathbf{1865921}}{3} \cdot 10^{54} \, .$$

where we have ten true decimal digits

$$\pi = 3.141\ 592\ 653\ 5\ldots,$$

which case exemplifies the standard first order approximation accuracy of formulae and equations in absolute geometry. The integer 1865921 has a meaning, too, but we leave it for later.

As one can infer, Euler's famous trigonometric formula extends into whatever possible super and ultra operators of geometry such as

(13) $$\left\{ e^{(2)\pi i} = \pm 1 \right\} \to e^{xyz\ldots\Phi \pi i_1} \to e^{5\Phi \pi i_1 e\sqrt{2}} \, .$$

What happens is that physics returns back to the real number domain. Moreover, physics begins to tie up strange knots generating natural numbers with definitely specific properties as above. Those whole numbers we will call **harmonious integers (HI).**

## Velocity of Light

The line that is straightest
offers most resistance.
**Leonardo da Vinci**

We have an obvious

(14) $$\frac{\text{Absolutes space}}{\text{Absolute time}} = \frac{\Phi}{i_1} = \text{Absolute velocity} = c_{Maxwell}$$

and also in terms of absolute Φ-dimensionality

(15) $$dim_{\Phi} c_{Maxwell} = i_1 .$$

Since 1984 it is accepted that, by definition,

$$c = 299792457 ms^{-1} .$$

Yet, because of the correlation of measurement units the non-dimensional number, 299792458, must be a theoretical number in depth. Indeed,

(16) $$c^{\pi} = \mathbf{427285} \cdot 10^{21} .$$

If accurately,

(17) $$c = 299792458 + \frac{1}{5.0676268180...} .$$

Mathematical symmetries here are broken due to the self-perturbation effects of geometry which are always explainable and computible in this geometry.

On the Picture A is seen the angle needed to start the Golden algorithm

(18) $$\arctan \frac{1}{2} = \Theta .$$

In geometry there is nothing else than Golden algorithm and, therefore,

(19) $$\sqrt[\sin\Theta]{299792457.975965378260...i_1} = \frac{10^{26}}{\mathbf{6480334}} .$$

As the reader learns geometry and gains experience, he/she will soon become able to calculate any possible violations of symmetries or prove the validity of any given HI.

# The Geometrizing God and the Standard Model

How strange it would be if the final
theory were to be discovered
in our own lifetimes!
**Steven Weinberg**

The greatest idea of the Western culture is neither geometry, nor Christianity, but the geometrizing God sweating over the problem of the Creation Act. What we aim at is to reconstruct on paper the compass and straightedge algorithm of Creaton.

The modern physics stops on what is called the Standard Model where the final result is the electroweak angle to be measured in precision experiments. We in turn dare to postulate that

$$\arctan\frac{1}{2} = \Theta_{Weinberg}. \tag{20}$$

In this case the electroweak force superoperator will be

$$e^{\Phi\pi\cdot\Theta_{Weinberg}} = \frac{10^{65}}{\mathbf{2263595}}. \tag{21}$$

We are imposed to face **the Universal Mathematical machinery** which we name the physical Universe. Workings of this machine is simple but subtle.

Interested readers may derive the following exponential growth operator by iterations

$$\exp\exp\chi = \chi\cdot 10^{68}. \tag{22}$$

Having gained certain experience in the logic and mathematics of absolute geometry, anyone will be in condition to foresee that

$$\sqrt[\chi]{e^{\Phi\pi i_1\cdot\Theta_{Weinberg}}} = \Phi^3 i_1\cdot 10^{14.00000...}. \tag{23}$$

So, as a mathematical fact, the postulate (20) looks consistent providing beautiful formulae. Then, how much is the probability that the geometrizing God unconsciously missed or consciously overlooked the aesthetics of mathematics in doing physics?

**"And thus much concerning God; to discourse of whom from the appearances of things, does certainly belong to Natural Philoshopy"**, said Sir Isaac Newton who founded physics as a theoretical discipline.

## The X-shaped Internal Geometry of the Point-Singularity

**Nihil in Verba**

University students of physics, at least, in my time were told that the electron is a structureless particle. As for me, I could never grasp this idea.

On the **Picture A** we see the space and time vectors. The picture still undergoes yet more spontaneous bifurcations. As a result, the intersection of the 4-Dim space-time point-singularity by a plane will reveal that the space and time vectors are intercrossed to construct either the straigth angle (photon) or the angle inverse to the experimental fine structure constant of atomic spectra (fundamental fermions). In the latter case we have roughly the **Picture B**, where

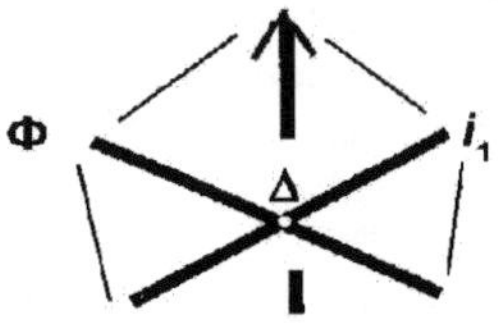


$$\frac{1}{\alpha_{Sommerfeld}} = \Delta_{Exprm.} = 137.035999... \tag{24}$$

Thus, most elementarily with respect to spinor particles

$$\left\{\Phi\cdot\Delta_{Exprm.}\cdot i_1\right\}^{\Phi\pi e} = \frac{\mathbf{64816}}{9}\cdot 10^{30}, \tag{25}$$

whence pure-theoretically

$$\frac{1}{\alpha_{Sommerfeld}} = \Delta_{Exprm.} = 137.0359990137371... \tag{26}$$

The most simple representation for the matter-radiation problem will evidently be a foreseeable

$$\sqrt[e]{\left\{\Phi\cdot\perp\Delta_{Exprm.}\cdot i_1\right\}^{\Phi\pi i_1}} = \mathbf{3}\cdot 10^{10}. \tag{27}$$

It is not difficult to guess also an exact

$$\left\{\Phi\cdot\perp\Delta_{Exprm.}\cdot i_1\right\} = \frac{1}{\chi}\sqrt[\chi]{\frac{\mathbf{2231071}}{3}\cdot 10^{20}}\text{, etc.} \tag{28}$$

This inner geometry of the nonlocal point does permanently oscillate between the two different states subject to the nonstop process of mutual bifurcations. In case of (X)-shape the point is the electron, while the (+)-shape is the photon described by

$$\left\{\Phi\perp i_1\right\}c = \frac{10^{12.9998...}}{\mathbf{180}}. \tag{29}$$

And the world is not without mystique, for the history of the Euclidean geometry has the following code (30):

$$^{Euclid}\mathbf{360}\cdot^{Gauss}\{\mathbf{3\cdot 5\cdot 17\cdot 257\cdot 65537}\}\cdot^{Post-Euclidean}\Phi\pi i_1 = 10^{12.9998...}.$$

And, for example,

(31) $$\sqrt{\frac{360\cdot\{3\cdot 5\cdot 17\cdot 257\cdot 65537\}\cdot\Phi\pi i_1}{\mathbf{180}\cdot\{\Phi\perp i_1\}}}\cdot i_1 = e^{10}.$$

## The *X*-constant of Universal Harmony

Tanto Monta
**Alexander of Macedonia**

The unified field theory (allow me a neologism: **unifield theory**) proves that the physical Universe we observe is the mathematical continuum self-excited and self-organized into **the universal system of harmony**. The key fact in the system is the tri-unity formula (2)

Then, the existing quantum-relativistic physics can be put as

(32) $$\frac{X\cdot(\alpha a_e)}{10} = 1+\varepsilon.$$

The Sommerfeld dimensionless constant, $\alpha$, is an anomaly, too, which case is clear from the classical quantum-mechanical

(33) $$\alpha = \frac{1}{4\pi\varepsilon_0}\cdot\frac{e^2}{\hbar c} = \frac{\mu_0 c e^2}{2h}.$$

There are no anomalies in Nature; anomalies arise because of discordance between the existing theories and objective reality. We have already seen that the problem concerns absolute dimensions of geometry.

In (32 ) mathematical symmetries are broken owing to the self-perturbation effect of geometry defined by

(34) $$\varepsilon = \frac{e^{2\Phi\pi i_1}}{\mathbf{131837531}}.$$

The HI on the right comes from a more deep lying reality

(35) $$\pi\frac{\Phi^3\sqrt{i_1 i_2}}{\text{Spin}}\cdot\mathbf{131837531} = \frac{\mathbf{23204152000}}{9},$$

where the entity on the left will be explained later. Here the accuracy of the first order approximation is

$$\pi = 3.141\ 592\ 653...,$$

while higher order approximations are always possible with the use of standard technique of absolute geometry.

## The Problem of Motion in Geometry

Most of our problems are test questions.
**Henry S. Haskins**

The milleniums long lasting problem of the true geometry of space-time is a cosmological test for intelligence which we are obliged to pass: otherwise, survival of human civilization on the cosmological scene will be impossible. The (+)-shaped photon will translate at the speed of light because of permanent bifurcation-oscillations of space and time vectors. This motion of a space-time point we name electromagnetic waves. So, we have originally and eventually a fundamental bifurcation such that

(36) $$\left[Space \Leftrightarrow Time\right] \rightleftarrows \left[Electrocity \Leftrightarrow magnetism\right],$$

which clarifies many mysteries in physics.

Thus, the space-time structure of electromagnetic waves is

(37) $$\left\{\Phi \perp i_1\right\}.$$

## The Quantum Gyroscope

Beauty is the first test.
**G. H. Hardy**

The X-shape defined by the Delta angle would look like as below on the **Picture C**. Such is the (sub)quantum space-time gyroscope and it can be somewhat computed in classical mechanics. Calculations definitely show that it has a mass

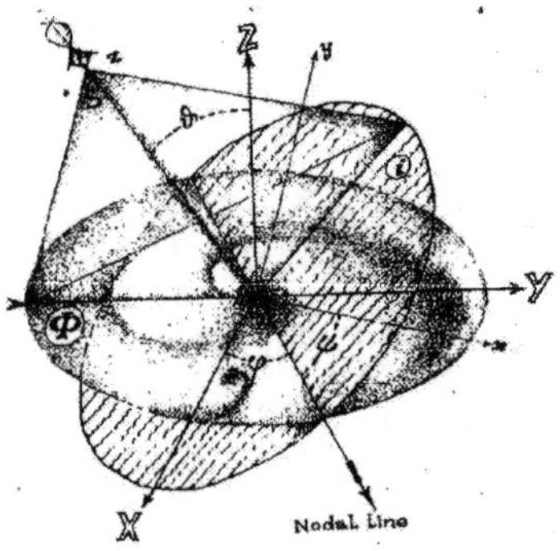


(38) $$m = \sqrt{2} \cdot$$

Consequently, the absolute dimension of, or else, the prototype for mass will be

(39) $$\dim m = \sqrt{2} \cdot$$

This gyroscope will translate and rotate and acquire any possible physical properties known or unknown to us.

Definition of the mass dimension in geometry is a most essential moment, for now we are enabled to write

(40) $$\dim G_{Newton} = \frac{\Phi^2}{\sqrt{2}}.$$

It should be always kept in mind that: The quantum gyroscope still oscillates between the straight angle and the kinematical Delta parameter and remains in the general form (27) written before.

Secondly, we already know samples of standard ways of doing things in geometry. Therefore,

(41) $$\sqrt[e]{\left\{G\dim G\left\{\Phi\cdot\perp\Delta_{Exprm.}\cdot i_1\right\}\cdot\dim m\right\}^{2\Phi\pi i_1}}=\frac{10^{33}}{\mathbf{1367182}},$$

deriving Newton's gravitational constant which does infinitely tend to the finite fraction: **6.673**. The integer on the right is, of course, not accidental.

Now be careful: We will try and derive the real experimental mass of the electron in the SI unit:

(42) $$\left\{G\dim G\left\{\Phi\cdot\perp\Delta_{Exprm.}\cdot i_1\right\}\cdot m_e\dim m\right\}=\mathbf{4039652}.$$

Whence,

(43) $$m_e=9.109381877$$

in excellent agreement with the empirical data.

The reader is already able to find out that

(44) $$\mathbf{4039652}^{\Phi\pi}=\exp\exp e\cdot10^{27},$$

which describes how things and phenomena do grow exponentially out of nothing. The Universe is a thought process, an abstraction. The fundamental mechanism is defined by a formula by definition

(45) $$1000\Phi\equiv e^{ee},$$

which says that the universal Fi-invariance of geometry and the exponential growth process are equivalent concepts.

What is immediately noticeable

(46) $$\{\Phi\perp i_1\}\cdot G\dim G=1000\dim E,$$

where we have in absolute dimensions

$$\dim(mc^2)=\Phi\sqrt{2}.$$

And also

(47) $$\{\Phi\perp i_1\}\cdot G\dim G\cdot\dim m=2000\Phi=\frac{2000}{\cos 72},$$

deriving the foundations of gravity in the simplest possible way.

## Bifurcation of Newton's Time

Time is the greatest innovator.
**Francis Bacon**

The unified theory is a fairly experimental science with the only difference that we are engaged in numerical experimentations. Observations unequivocally show that Newton's time is bifurcated

(48) $$\left\{i_1 = \sqrt{\Phi}\right\} \leftrightarrow \left\{i_2 = \frac{4}{\pi}\right\}.$$

Intuitively, it is clear from the beginning that

(49) $$\left\{\pi \leftrightarrow \Delta_{137}\right\}$$

in terms of curvature and who knows if by such a curious reason that

(50) $$\pi = 3\frac{1}{7}.$$

This kind of speculations leads to a discovery of

(51) $$\sqrt[\sin \Delta_{Exprm.}]{\left\{\frac{4}{\pi}\right\}^{\Delta_{Exprm.}}} = \mathbf{124201} \cdot 10^{16},$$

showing a close relationship between the Pi and Delta.

But, Newton's time bifurcates not because we know the above or we can draw in this geometry anything like (Picture D)

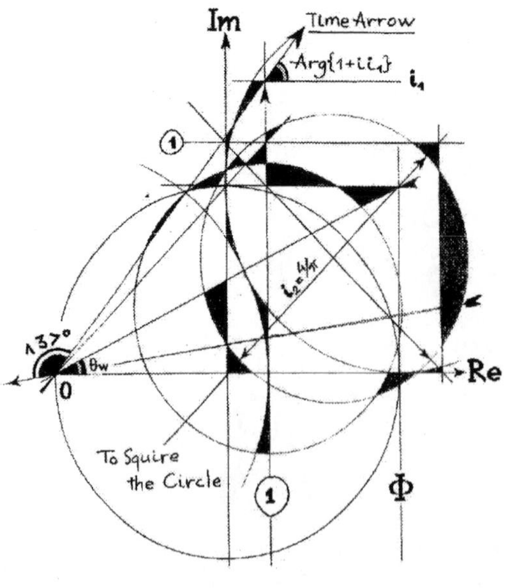

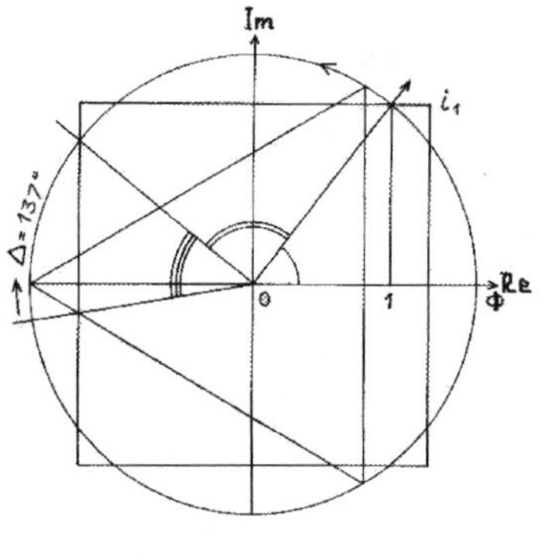


This drawing (Picture E) by Ts. Taikhar explains the algorithmic existence of the Delta angle. In any event, we are left to consider the cosmological average

(52) $$\sqrt{i_1 i_2}\,.$$

Consequently, the 4-Dim space-time has to be written as

(53) $$\Phi^3\sqrt{i_1 i_2}\,.$$

If make it complete with curvature and torsion, then the point-singularity of geometry shall be

(54) $$\pi\frac{\Phi^3\sqrt{i_1 i_2}}{\text{Spin}}.$$

And this is always valid, for example,

(55) $$\left\{\pi\frac{\Phi^3\sqrt{i_1 i_2}}{\text{Spin}}\right\}^{G\dim G}=\cos\Theta_W\cdot 10^{16},$$

showing gravity as a golden-algorithmic phenomenon in depth.

As for the dimensionality of the Φ-invariant space-time,

(56) $$\dim_\Phi \Phi^3 i_1=\dim_\Phi \Phi^3\sqrt{i_1 i_2}=3\frac{1}{2}$$

and in Cartesian terms

(57) (3+1).

And, all this satisfies the following Golden-algorithmic equation:

(58) $$\pi\frac{\Phi^3\sqrt{i_1 i_2}}{\text{Spin}}\cdot\left[\dim_\Phi \Phi^3 i_1\cdot(3+1)\right]\cdot$$
$$\cdot X\cdot e^{5\Phi\pi e i_1\sqrt{2}}\cdot e^{8\Phi\pi i_1}=\cos\Theta_W\cdot 10^{85}.$$

Also, it is foreseeable that

(59) $$\frac{\pi\frac{\Phi^3\sqrt{i_1 i_2}}{\text{Spin}}\cdot\dim_\Phi \Phi^3 i_1(3+1)}{\alpha a_e\{1+\varepsilon\}}=2\Phi\cdot 10^7\,;$$

(60) $$\sqrt[e]{\left\{\frac{1}{\varepsilon}\right\}^{\pi}}=\frac{\mathbf{48848}}{3}.$$

## Quantum Leap Operators

Your purpose in life is simply to
help on the purpose of the universe.
**George Bernard Shaw**

Geometry needs in quantum leap operators such as

(61)
$$e^{©} = ©^{-1};$$
$$\ln © = -©;$$
$$© = 0.567149290405...$$

One can calculate yet another obvious one by iterations

(62)
$$\lg @ = -@;$$
$$10^{@} = @^{-1}.$$

Besides,

(63)
$$ABCdef... = \frac{1}{CBAdef...};$$
$$\exists = \mathbf{571100522647...};$$
$$\frac{1}{\exists} = \frac{1}{\mathbf{175}100522647...}.$$

Its most usable decimal form is

(64) $$\exists = 5.71100522647...$$

Given the following obvious configuration of these operators

(65) $$\frac{\exists}{©@},$$

we will obtain a sample of the quantum jump Bigbang cosmological scenario

(66) $$\frac{\exists}{©@} \cdot \pi \frac{\Phi^3 \sqrt{i_1 i_2}}{\text{Spin}} = \sqrt[\chi]{\frac{\mathbf{13052875}}{3} \cdot 10^7};$$

(67) $$\mathbf{13052875}^{\pi} = ©@ \cdot 10^{23}.$$

The inversed-E operator can be the largest whole number or the smallest decimal fraction at one and same time depending on circumstances both in Nature and in theories.

## The Big Bang Scenario of Cosmology

Nothing puzzles me more than time and space; and yet nothing puzzles me less, for I never think about them.

**Charles Lamb**

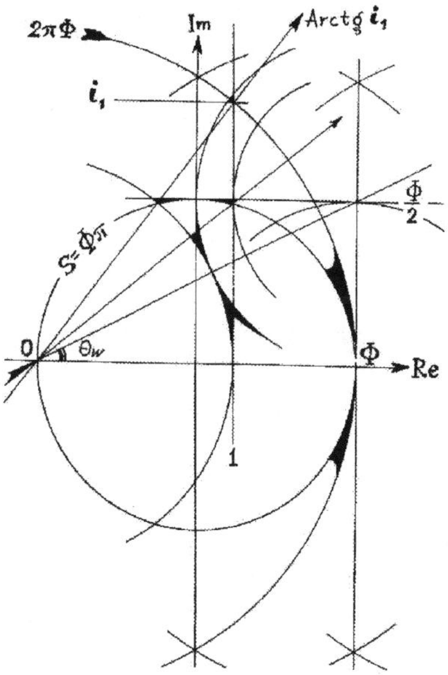


The plane absolute geometry is seen on the Picture F. In this introductory book we omit incredibly more complicated drawings designed to be mathematical machines for computing fundamental constants. It will be a good game for interested students.

An approach to the Bigbang scenario will certainly be

$$\text{(68)} \qquad \sqrt[©@]{\pi \frac{\Phi^3 \sqrt[3]{i_1 i_2}}{\text{Spin}}} = \Phi\pi \cdot 10^5 .$$

What we have done up till the moment is the arithmetization of the Universe, according to the Pythagoras' concept. To an extent, Pythagoras foresaw the key importance of the golden constant for the explanation of the world.

Now it is time to move to the Plato's concept of geometrization. Plato saw foundations of matter in the regular figures of the Euclidean geometry. However, at this point arises a question: How it is that the self-excited and self-organizaed mathematical continuum (software) is observed and perceived by all our senses as something material (hardware)?

To eliminate controversies between the material and the ideal, we are required to postulate the existence of topological densities such that

$$\text{(69)} \qquad (x+y+z+\ldots)(x\times y\times z\times\ldots) = \{\mathbf{x},\mathbf{y},\mathbf{z},\ldots\}_+^\times .$$

In case of regular polygons we have to have

$$\{\mathbf{3},\mathbf{5},\mathbf{17},\mathbf{257},\mathbf{65537}\}_+^\times ,$$

which satisfies

$$\text{(70)} \qquad \sqrt[\pi]{\{\mathbf{3},\mathbf{5},\mathbf{17},\mathbf{257},\mathbf{65537}\}_+^{\times 5}} = 10^{23} .$$

In the spirit of Pythagoras and Plato we derive Newton's constant by a most standard trick of the absolute geometry as follows:

$$(71)\qquad \sqrt[e]{\left\{G\dim G\{\mathbf{3,5,17,257,65537}\}_{+}^{\times}\right\}^{2\Phi\pi i_1}} = \frac{\mathbf{2638453}}{3}\cdot 10^{68}.$$

The 3-Dim Plato’s solids are treated the same way, but we leave them and move immediately to the 4-Dim polytops whose topological density composed of their topological parameters will be

$$(72)\qquad \mathbf{D}\{4\}_{+}^{\times} = \begin{array}{|c|c|c|c|l} \hline \mathbf{5} & \mathbf{10} & \mathbf{10} & \mathbf{5} & \times \\ \hline \mathbf{8} & \mathbf{24} & \mathbf{32} & \mathbf{16} & \\ \hline \mathbf{16} & \mathbf{32} & \mathbf{24} & \mathbf{8} & \\ \hline \mathbf{24} & \mathbf{96} & \mathbf{96} & \mathbf{24} & \\ \hline \mathbf{120} & \mathbf{720} & \mathbf{1200} & \mathbf{600} & \\ \hline \mathbf{600} & \mathbf{1200} & \mathbf{720} & \mathbf{120} & \\ \hline \end{array}_{+} .$$

It is an astronomically large number

$$(73)\qquad \mathbf{D}\{4\}_{+}^{\times} = \mathbf{2,83384389862590703418380124 16e+45}.$$

Yet, simple but beautiful a logos

$$(74)\qquad \exists = \frac{\Phi}{\mathbf{D}\{4\}_{+}^{\times}}\left\{1-\sqrt{\frac{\Phi^{3}\sqrt{i_1 i_2}}{10^{8}}}\right\}.$$

Besides, geometry often ends up with expressions of the type

$$(75)\qquad \arccos\lg\lg\lg\lg\sqrt[e]{\mathbf{D}\{4\}_{+}^{\times\Phi\pi}} = \Phi^{10},$$

highlighting the problem of these two radices in the absolute calculus.

Now that the concisest possible description of the quantum leap Bigbang origin of the Pythagoras-Plato 4-dimensional space-time world shall be as follows:

$$(76)\qquad \frac{10^{\mathbf{46}}}{©} = \frac{\Phi^{3} i_1}{\text{Spin}}\cdot\mathbf{D}\{4\}_{+}^{\times}.$$

And this world will self-gravitate

$$(77)\qquad G\dim G\left\{\sqrt{\pi\frac{\Phi^{3}\sqrt{i_1 i_2}}{\text{Spin}}}\cdot\mathbf{D}\{4\}_{+}^{\times}\right\} = \mathbf{6846072}\cdot 10^{41}.$$

## The Problem of Absolute Dimensions

No way of thinking or doing, however ancient, can be trusted without proof.
**Henry David Thoreau**

The tri-unity of basic dimensions are:

(78) $$^{Space}\Phi, \quad ^{Time}i_1, \quad ^{Mass}\sqrt{2}$$

and they are fully embraced by the Planck quantum constant of action:

(79) $$\dim h = \Phi i_1 \sqrt{2}\,.$$

We should, of course, write a topological density

(80) $$\{\Phi, i_1, \sqrt{2}\}_{+}^{\times}.$$

Now the reader is asked to compose the topodensity of Plato's 3-dimensional solids and subsequently derive the following cosmic space-time configuration, by definition:

(81) $$\{\Phi, i_1, \sqrt{2}\}_{+}^{\times} \cdot \left\{ \pi \frac{\Phi^3 \sqrt{i_1 i_2}}{\text{Spin}} \cdot \dim_{\Phi} \Phi^3 i_1 (3+1) \right\} \cdot$$
$$\cdot \mathbf{D}\{\mathbf{2}\}_{+}^{\times} \cdot \mathbf{D}\{\mathbf{3}\}_{+}^{\times} \cdot \mathbf{D}\{\mathbf{4}\}_{+}^{\times} \equiv \frac{10^{78}}{a_e}.$$

Here and elsewhere 'by definition' will mean that the given expessions are true in principle and can be approximated to whatever extent by methods intrinsic to the absolute geometry alone. In the given case

(82) $$\ldots = \frac{10^{78}}{a_e}\{1-\varepsilon\}; \qquad \frac{1}{\varepsilon} \cdot e^{3\Phi\pi i_1} = \frac{\mathbf{7284776}}{9} \cdot 10^6.$$

What does this tell us? The electron magnetic moment anomaly, or else, the post-Dirac imprecision of quantum-relativistic theories, owes to the self-perturbation effect of the entire geometry.

Thanks to the universality of the system under discussion, any of fundamental physical constants (coefficients) can and must be derived in many ways intrinsic to the absolute geometry, including quite non-trivial ones

(82) $$\{\Phi, i_1, \sqrt{2}\}_{+}^{\times} \cdot h = \sqrt[\chi]{10^e \cdot 10^7}\,;$$

(83) $$\{\Phi, i_1, \sqrt{2}\}_{+}^{\times} \cdot h = \sqrt[\Phi\pi e i_1]{\Phi^3 i_1 \cdot 10^{33}}\,.$$

Bare numeric coefficients arise by whatever overlapping equivalent reasons of mathematical garmony, for example,

(84, 85) $$G^{\otimes} = \frac{999.999}{\pi - 3}; \qquad h^{\oplus} = \tan\frac{2\pi}{5} \cdot 10^{6.99999...};$$

(86, 87) $$10^{\otimes} = 10000 \otimes; \qquad e^{\oplus} = 1000 \oplus \cdot$$

This kind of game can be unfolded to whatever extent:

(88) $$G^{\otimes} h^{\oplus} = \frac{10^{13}}{\mathbf{46}};$$

(89) $$G^{\otimes} h^{\oplus} \cdot e^{\Phi \pi i_1} e^{5\Phi \pi i_1 e \sqrt{2}} = \mathbf{132009} \cdot 10^{63} \cdot$$

## Mathematical Cosmology

I had a feeling once about Mathematics - that I saw it all.

**Winston S. Churchill**

What we have been trying to study in this book is, in fact, mathematical cosmology. The quantum leap process resulted in the Pythagoras-Plato 4-world such that

(90) $$\left\{\{\Phi, i_1, \sqrt{2}\}_+^\times \cdot h\right\} \left\{\pi \frac{\Phi^3 \sqrt{i_1 i_2}}{\text{Spin}} \cdot \dim_\Phi \Phi^3 i_1 (3+1)\right\} \cdot$$
$$\cdot \mathbf{D}\{\mathbf{2}\}_+^\times \cdot \mathbf{D}\{\mathbf{3}\}_+^\times \cdot \mathbf{D}\{\mathbf{4}\}_+^\times \equiv \exists.$$

If accuracy needed, then correct the right section by

(91) $$1 + \sqrt[\pi e]{\frac{1}{\pi \Phi^3 i_1 \cdot 10^{31}}} \cdot$$

So far so good; we have picked up a mathematical point and blown it up into the whole Universe. And this Bigbang scenario does not depend on any prior knowledge of the current physical cosmology. The minimax principle of logic will allow only the above scenario.

The modern physical cosmology looks quite a convincing scheme grounded mainly on the observational data rather than on fundamental theories. If formally, the unified theory proves the modern cosmology. At the same time, the unifield theory does not rule out the possibility that the observational data might have been misinterpreted. But, this is too subtle a problem we have to entirely leave in this introductory book.

## Quantum Gravitation Solution

Exercize, exercize your powers: what is now difficult will finally become routine.
**G. C. Lichtenberg**

In the Creation act God made few, but most graceful allowances to us, fools. It is, first, Newton's coefficient tending to a finite fraction. This eases the whole work on the unifield theory. The second one is even more favourable:

(92) $$\dim(Gh) = \Phi^3 i_1$$

implying that the 4-spacetime is a quantum-gravitational phenomenon from the beginnig. Therefore, the problem of quantum gravity said to be damn difficult becomes simply cancelled by the $\Phi$-invariance. There is no such problem as quantum gravity at all.

Having previously learned particularities of space-time, we guess to try and find that

(93) $$\sqrt[e]{\left\{\frac{Gh\cdot\Phi^3 i_1}{\alpha a_e}\right\}^{2\Phi\pi i_1}} = \frac{10^{41}}{\mathbf{1\cdot2\cdot3\cdot4\cdot5\cdot6\cdot7\cdot8\cdot9}},$$

reducing the problem to the configuration of arithmetic.

We have seen that

$$G\dim G\cdot\left\{\pi\frac{\Phi^3\sqrt{i_1 i_2}}{\text{Spin}}\cdot\mathbf{D\{4\}}_{+}^{\times}\right\} = \mathbf{6846072}\cdot10^{41}.$$

The process ends up with the reference to the Sancta Sanctorum of geometry to be explained later :

(94) $$\mathbf{6846072}\cdot2\Theta_{STR.} = \mathbf{457796134}.$$

The quantum-gravitational cosmic vacuum has to be written rather in tri-unity form, and there is already something

(95) $$Ghc = \frac{\mathbf{1193}}{9}.$$

Indeed,

(96) $$\sqrt[\Phi]{\{Ghc\cdot\dim(Ghc)\}^{\pi e}} = \frac{\mathbf{1232617}}{3}\cdot10^{10}.$$

This notably proves that Newton's and Maxwell's constants can and must be put by definition as equal to 6.673 and 2.99792458, correspondingly, so that to define the theoretical value of quantum constant as next

(97) $$h = 6.626068761\cdot$$

The previous statement will be true if only the HI on the right occurs by some fundamental reasons. Indeed, most consistently

(98) $$\mathbf{1232617} \cdot \widehat{E}\breve{E} = \pi \cdot 10^{28},$$

where the left section includes the **energy-entropy configuration** of geometry to be derived later.

After all, it is simply that

(99) $$\{Ghc \cdot \dim(Ghc)\} \cdot \mathbf{U}_{\mathbf{E}} = \frac{\mathbf{18482}}{9} \cdot 10^{9},$$

where on the left is what we will name **universal energy** compared with Einstein's conventional energy.

The *Ghc*-triad is fundamental owing to

(100) $$\mathbf{18482} \cdot e^{\Phi \pi i_1} \equiv \tilde{\mathbf{i}} \cdot 10^{7},$$

where we have a most mystical looking operator

(101) $$\tilde{\mathbf{i}} = \left\{ \frac{1}{\cos i - 1} - 1 \right\}^{-1},$$

though the logic is simply that the Being results as if from the multiply inversed, or bifurcated, operator of Nonbeing denoted as

$$-1$$

We have in the above introduced three things so far unexplained to the reader, notably,

(102) $$\cos 2\Theta_{STR.}, \quad \widehat{E}\breve{E}, \quad \mathbf{U}_{\mathbf{E}}.$$

By the moment the theory has become easy, while systematical rendition of the theory poses a hard task. For example, what we call the energy-entropy configuration is the result of colossal number of observations and experimentations that took years. It looks something chaotic and unbelievable until after the reader repeats all the observations and experimentations which case is hardly possible in practice. So, for a while the reader is supposed to trust the author.

Whatsoever, the above triad as a whole and each of them in separate is ultimately logical and consistent in the system of absolute geometry, for

(103) $$\cos 2\Theta_{STR.} \times \widehat{E}\breve{E} \times \mathbf{U}_{\mathbf{E}} = ©@ \cdot 10^{32}.$$

## The Golden Braid

Nothing hurts a new truth
more than an old error.
**Goethe**

The previous texts' purpose was just to show to the reader that the Φ-invariance of geometry is going to be an efficient solution for the ultimate synthesis of all the scientific knowledge obtained earlier in history by the analytical methods, the latter being based on what is called applied mathematics founded by Newton and Leibniz.

This time science returns back to the pure mathematics. It is a shocking experience to us that the most fundamental laws of Nature appear to be written in the language of natural numbers. Therefore, mathematical mythology turns out to be true.

The final theory of physics occur to be a pure mathematical one and it belongs to the theory of numbers. Is it too strange to be trusted? No, not so strange, as it would seem to physicists. According to Kurt Goedel's theorem, any possible final theory of physics would, or rather should, belong to a higher system than physics proper.

Fundamental physics looks now infinitely simplified. Calculation of the formula (10) takes seconds and the formula itself is consistent in the absolute-geometric system. As it is reported recently, the same calculation of the anomaly of the electron magnetic moment took Toichiro Kinoshita and his group more than 10 years of work on supercomputers.

Simplification of physics implies democracy in the field of fundamental research. Everyone wishful will feel free to experiment with the SUT, discover something new in mathematics and physics and trade their own results in science and technology.

The absolute geometry is Golden-algorithmic offering the shortest possible road to the cosmological reality. The following suffices to around up with the modern physics:

(104) $$\sqrt[\sin\Theta_W]{\frac{1}{a_e}} = \mathbf{3667438};$$

(105) $$\sqrt[\sin\Theta_W]{\frac{1}{\alpha a_e}} = \frac{9}{\mathbf{409046}} \cdot 10^{16}.$$

The Golden section was a leading cognitive ideology during the ancient Sumerian civilization, and the antique Greek ages, and the European Renaissance. This time it professes the global new progress of humanity.

The problem of quantum gravity was the hardest nut for the contemporary phyiscs, notably, because of its fundamentality and simplicity. It conceils behind the following golden braid of harmonious integers:

(106) $$\sqrt[\sin\Theta_W]{\frac{Gh\cdot(\dim Gh=\Phi^3 i_1)}{\alpha a_e}}=\frac{\mathbf{4091872}}{9}\cdot 10^{11};$$

(107) $$\sqrt[e]{\mathbf{4091872}^{\Phi\pi i_1}}=\frac{\mathbf{16025}}{3}\cdot 10^{12};$$

(108) $$\sqrt[3+1]{\sqrt[\dim_\Phi \Phi^3 i_1]{\mathbf{16025}^{\pi\frac{\Phi^3 i_1}{\text{Spin}}}}}=\mathbf{742580};$$

(109) $$\mathbf{742580}\cdot e^{5\Phi\pi e i_1\sqrt{2}}\cdot\widehat{E}\breve{E}=\mathbf{1788}\cdot 10^{79};$$

(110) $$\mathbf{1788}^{\chi}=\mathbf{29477526}\cdot 10^{9};$$

(111) $$\sqrt[e]{\mathbf{29477526}^{2\Phi\pi i_1}}=\frac{9}{\mathbf{2624924}}\cdot 10^{41},$$

where the last one is accurate with

$$\pi=3.141\ 592\ 653\ 58...$$

It should be remarked that since antique times we have the holy trinity of fundamental concepts of:

**Universal invariance** (Thales);
**Arithmetization** (Pythagoras);
**Geometrization** (Plato).

Then, how to formulate an ultimate possible theorem? This genial job appears to have been done as early as by Giordano Bruno in 1591.

The Bruno's theorem reads:

**The order of a unique figure and**
**the harmony of a unique number**
**give rise to all things.**

Needless to say that the superunified theory is no more and no less than the long delayed proof of this metatheorem of the natural sciences.

## Constants and Operators

Everything that is created
Is part of a mutual order, and that is the shape
Which makes the universe resemble God.
Here the superior beings see the traces
Of the eternal power...

**Dante Aligieri**
The Divine Comedy
Paradiso I, 103/106

To ease the study of the unified theory, the constants and operators of geometry, mostly newly discovered, has to be memorized.

To work in confidence and good faith, compute first a golden-section superoperator

(112) $$e^{\pi i_1 \cdot \Theta_W \sin\Theta_W} = \mathbf{4152773} \cdot 10^{14};$$

(113) $$\{\mathbf{4,1,5,2,7,7,3}\}_{+}^{\times(\pi e)} = \frac{9}{\mathbf{1889173}} \cdot 10^{50}.$$

The universal four-dimensional space-time (or else, the point-singularity) with curvature and torsion is

(114) $$\mathrm{Sing} = \pi \frac{\Phi\sqrt{i_1 i_2}}{\mathrm{Spin}},$$

where according to quantum mechanics

$$\mathrm{Spin} = \cos 30°.$$

And

(115) $$\dim_\Phi \mathrm{Sing} = 3\tfrac{1}{2};$$

(116) $$\dim_{Cartesian} \mathrm{Sing} = 3+1.$$

The hyperbolic cosine

$$\cos i = 1.54308063481524377847790...$$

works in any possible form, including

(117) $$\cos\tilde{\mathbf{i}} = \left(\frac{1}{\cos i - 1} - 1\right)^{-1} = 1{,}1885699670...;$$

(118) $$\pi\frac{\Phi^3\sqrt{i_1 i_2}}{\mathrm{Spin}} \cdot \dim_\Phi \Phi^3 i_1 \cdot (3+1) = \sqrt[\cos^3 i]{\mathbf{902824000}}.$$

Constants of (sub)quantum leap processes are

(119, 120)
$$© = 0.5671432904097838729999686...;$$
$$@ = 0.3990129782602520715964708 1...,$$

(121) $\ln © = -©$; $e^{©} = ©^{-1}$;

(122) $\lg @ = -@$; $10^{@} = @^{-1}$;

(123) $©@ = 0{,}2262975334067269197232763983607$;

$$\exists = 571\ 100\ 522\ 647\ 717\ 554\ 761\ 906\ 724\ 501...;$$

(124) $\frac{1}{\exists} = 175\ 100\ 522\ 647\ 717\ 554\ 761\ 906\ 724\ 501...$

Samples of logarithmic operators:

(125) $\otimes_0 = 4.66924683287747670306999697 91...;$

$$10^{\otimes} = 10000 \otimes;$$

(126) $\oplus = 9.118006470402740121258337182046 81...;$

$$e^{\oplus} = 1000 \oplus \cdot$$

(127) $\chi = 5{,}0638468616142932 0....;$

$$\exp e^{\chi} = \chi \cdot 10^{68}.$$

Delta operators as related to the kinematics of energy are:

(128) $\Delta_0 = \frac{1}{0.0072};$

$$\Delta_1 = \frac{360}{\Phi^2} = 137.5077640500...^{\circ};$$

$$\Delta_{\alpha} = 137.0916934^{\circ};$$

$$\Delta_{\beta} = 137.0405966^{\circ};$$

$$\Delta_{Exprm.} = \frac{1}{\alpha_{Sommerfeld}} = 137.035999...;$$

(133) $\nabla = \frac{360}{\Phi} = 222.4922359^{\circ}.$

Parameters of the internal geometry of nonlocal point-singularity:

(134) $$\Theta_{Weinberg} = \arctan\frac{1}{2} = 26.5650511770°;$$

(135) $$\Theta_{Strong\ nuclear\ force} = \frac{\pi}{3} - \Theta_W = 33.4349488229220°;$$

(136) $$\cos 2\Theta_{STR.} = 0.392820323;$$

(137) $$ArgZ_1 = Arg\{1 + ii_1\} = 51.82729237°;$$

(138) $$ArgZ_2 = Arg\{\Phi + ii_1\} = 38.17270763°.$$

The general form of topological configurations:

(139) $$(x + y + z + ...)(xyz...) = \{\mathbf{x, y, z, ...}\}_{+}^{\times}.$$

Configurations of the internal geometry of singularity can be written in many different ways, including

(140) $$\mathbf{InG}_{+}^{\times} = \{\Delta_1, \nabla, \Theta_W, \Theta_{STR.}, ArgZ_1, ArgZ_2\}_{+}^{\times} = \\ = 27417955234218{,}634496061908573843;$$

(141) $$\mathbf{I/G} = \{\Delta_{Exprm.}, \Theta_W, \Theta_{STR.}, Arctgi_1, artctg_{i_1}^{-1}\}_{+}^{\times}.$$

Platonic configurations of regular topology (142-144):

$$\mathbf{D\{2\}}_{+}^{\times} = \{\mathbf{3,5,17,257,65537}\}_{+}^{\times} = 282690452389605;$$

$\mathbf{D\{3\}}_{+}^{\times} =$ (table, × at top right, + at bottom right)

| | | |
|---|---|---|
| 4 | 6 | 4 |
| 6 | 12 | 8 |
| 8 | 12 | 6 |
| 12 | 30 | 20 |
| 20 | 30 | 12 |

$= 313714645401600000$ ;

$\mathbf{D\{4\}}_{+}^{\times} =$ (table, × at top right, + at bottom right)

| | | | |
|---|---|---|---|
| 5 | 10 | 10 | 5 |
| 8 | 24 | 32 | 16 |
| 16 | 32 | 24 | 8 |
| 24 | 96 | 96 | 24 |
| 120 | 720 | 1200 | 600 |
| 600 | 1200 | 720 | 120 |

$$= 2.8338438986259070341838012416 \cdot 10^{45}.$$

One of possible compositions of global topology will be

(145) $$\{\mathbf{D\{2\}}_{+}^{\times},\ \mathbf{D\{3\}}_{+}^{\times},\ \mathbf{D\{4\}}_{+}^{\times}\}_{+}^{\times}\cdot\mathbf{InG}_{+}^{\times}\equiv\frac{1}{(©@)^2};$$

(146) $$\frac{1}{(©@)^2}=\lg(\mathbf{33668}\cdot 10^{15}).$$

In this work we make use of experimental coefficients of physical constants, neglecting experimental uncertainties:

(147-151)
$$c_{Maxwell}=2.99792458;$$
$$G_{Newton}=6.673;$$
$$h_{Planck}=6.62606876;$$
$$m_{electron}=9.10938188;$$
$$e^{\pm}_{Coulomb}=1.602176462.$$

Dimensions in absolute-geometric terms are defined by

(152) $$\mathbf{X}_{\textit{Numeric coeff.}}\times^{\textit{Prototype}}\mathbf{Dim}_{\Phi}\mathbf{X}=\text{Unified Configuration}.$$

Dimensions of universal physical constants are:

(153-158)
$$\mathrm{Dim}_{\Phi}G=G_{ag}=\frac{\Phi^2}{\sqrt{2}}=\frac{1}{\Pi_{\text{Cosmological}}};$$
$$\mathrm{Dim}h=h_{ag}=\Phi i_1\sqrt{2};$$
$$\mathrm{Dim}m=\sqrt{2};$$
$$\mathrm{Dim}e^{\pm}=\Phi\sqrt{\sqrt{2}};$$
$$\mathrm{Dim}c=i_1;$$
$$\mathrm{Dim}E=E_{ag}=\Phi\sqrt{2}.$$

Universal gravitational operators:

(159) $$G\dim G=GG_{ag}=6.673\frac{\Phi^2}{\sqrt{2}}=\frac{G}{\Pi_{Cosm.}}==12{,}353255032862...;$$

(160) $$\frac{G}{G_{ag}}=6.673\cdot\frac{\sqrt{2}}{\Phi^2}=G\Pi_{Cosm.}=3.604631239.$$

Energy configuration in Einstein's terms (161):

$$E=\left\{\left[N=\frac{1}{2}\left(\frac{m_{Proton}}{m_e}+\frac{m_{Neutron}}{m_e}\right)\right]\cdot m_e \dim m\right\}\cdot(ci_1)^2=$$

$$=\mathbf{344222,3208...}$$

Universal energy configuration:

(162) $$\mathbf{U_E}=E_{ag}\cdot hh_{ag}\cdot 2\mathrm{N}m_e c^2\cdot\frac{2\Theta_{STR.}}{\cos 2\Theta_{STR.}}=$$

$$=\mathbf{2260258601.268245380...}$$

Energy-entropy configuration:

(163) $$\widehat{E}\breve{E}=\frac{\exists}{©@}\cdot\pi\cos\frac{180}{\pi}\cdot 10^{\frac{1}{\dim E}}\cdot\left(\frac{\dim E}{2}\right)^{3+1}\cdot$$

$$\cdot k_{Boltzman}\cdot\mathbf{U_E}\cdot E\cdot\frac{\Delta_{Exprm.}}{a_e}.$$

Fundamental mass ratio:

(164) $$N=\mathbf{1837,41}$$

Electron magnetic moment anomaly:

(165) $$a_{Electron}=\mathbf{0.0011596521}=\frac{1}{\aleph_1}.$$

Topology of Feodorov's crystallography:

(166) $$\mathbf{Cr}_+^\times=\mathbf{\{2,13,59,68,36,52\}}_+^\times\cdot\mathbf{\{230,32,14,6\}}_+^\times=$$

$$=7.830223555\times 10^{18}.$$

Topology of Urmantsev's system morphisms:

(167) $$\mathbf{Ur}_+^\times=\mathbf{\{162,192,360,55584,8,255\}}_+^\times=$$

$$=7,181510295\cdot 10^{19}$$

Topology of gene code:

(168) $$\text{🕸}=\mathbf{\{1,1,5,2,9,2,6,3,9,5,9,}$$

$$\mathbf{5,6,5,6,5,4,1,3,20,61,3\}}_+^\times=\mathbf{4.434768857\times 10^{16}}.$$

Topology of human chromosome structure:

(169) $$\text{☺}=\mathbf{\{4,7,46,2,3,2,7,3,1,2,2,2,1\}}_+^\times=$$

$$=\mathbf{212921856.}$$

Topology of biochemistry:

(170)
$$\mathit{Chemy} = \\ = \{\mathbf{H}^1, \mathbf{O}^{16}, \mathbf{C}^{12}, \mathbf{N}^{14}, \mathbf{P}^{30}, \mathbf{S}^{32}\}_+^\times \cdot \{\mathbf{H}_2^2, \mathbf{O}^{16}\}_+^\times = \\ = 3.46816512 \cdot 10^{11}.$$

The above biology related configurations are written tentatively.

Distribution of human genes by chromosomes according to the table circulated online several years ago by the US Ministry of Enegy:

(171)

| | | |
|---|---|---|
| 1/**3148** | 9/**1229** | 17/**1367** |
| 2/**902** | 10/**1312** | 18/**365** |
| 3/**1436** | 11/**405** | 19/**1553** |
| 4/**453** | 12/**1330** | 20/**816** |
| 5/**609** | 13/**623** | 21/**446** |
| 6/**1585** | 14/**886** | 22/**595** . |
| 7/**1824** | 15/**676** | ***X*/1093** |
| 8/**781** | 16/**898** | ***Y*/125** |

In this regard there are circulated different data and the end result of the genome mapping project is not clear yet. The previous table looks quite consistent from the absolute-geometric point of view:

(172)
$$\sqrt[\Phi]{\{\mathbf{Genes}\}_+^\times} = \tan\frac{2\pi}{5} \cdot 10^{\mathbf{46}}$$

Besides, not just for fun

(173)
$$\{X_{1093}, Y_{125}\}_+^\times \cdot \frac{\mathbf{D}\{\mathbf{4}\}_+^\times}{\mathrm{Sing}} = \mathbf{24114} \cdot 10^{48};$$

(174)
$$\mathbf{24114} = \sqrt[\pi\pi]{\frac{\mathbf{5350387}}{3} \cdot 10^{37}}\,.$$

If rather just for fun than not, then the topology of human genetics would look like

(175)
$$\mathbf{Homo} = \mathit{Chemy} \cdot [\![\text{🕸}\text{😐}]\!] \cdot \{\mathbf{Genes}\}_+^\times = \mathbf{542891047} \cdot 10^{103};$$

(176)
$$\left\langle 542981047 \cdot 10^{103} \right\rangle^{\pi \frac{\Phi^3 i_1}{\mathrm{Spin}} \cdot \dim_\Phi \Phi^3 i_1} = \frac{10^{7645}}{2\pi}.$$

In this geometry are often nesessary calculations of the type

(177) $$\frac{1}{10}\cdot\sqrt{10\cdot\sqrt{arch\frac{360}{\lg\frac{\mathbf{7645}^{\Phi\pi i_1 e}}{180}}}}=\cos 2\Theta_{STR.}.$$

The reason is here a fundamental algebraic bifurcation

(178) $$\cos i \leftrightarrow \cos 2\Theta_{STR.}.$$

There are two or three points to be highlighted from the beginning. First, new theories tend to a bit exaggerate things; otherwise, it is impossible to investigate things deeper. Secondly, the final goal of cognition is man and the unified theory has to define man with every possible rigour; otherwise, there will be completely no need to make much ado about nothing. Thirdly, the central observational fact to be explained in theoretical sciences is the existence of an intelligent observer who we are.

And, however unbelievable the previous crazy speculations were, we nevertheless obtain a consistent representation (179)

$$\mathbf{542891047}\cdot\left\{\begin{matrix}\pi\frac{\Phi^3\sqrt{i_1 i_2}}{\text{Spin}}\cdot\dim_\Phi\Phi^3 i_1(3+1)\cdot\\ \cdot\frac{\mathbf{D\{2\}}_+^\times\cdot\mathbf{D\{3\}}_+^\times\cdot\mathbf{D\{4\}}_+^\times}{\mathbf{InG}_+^\times}\cdot\mathbf{U_E}\end{matrix}\right\}=\mathbf{3079426}\cdot 10^{78},$$

where on the left is the most standard global topological configuration of this geometry.

Toplogical configuration of arithmetic:

(180) $$\mathbf{\{1,2,3,4,5,6,7,8,9\}}_+^\times=\&_+^\times.$$

Compute immediately

(180) $$\mathbf{\&}_+^\times\cdot\frac{1}{\alpha a_e}=\mathbf{\&}_+^\times\cdot\Delta_{Exprm.}\aleph_1=\sqrt[\pi]{\frac{\mathbf{3549965}}{9}\cdot 10^{33}}.$$

A typical picture in this geometry is as follows:

(181) $$\Phi=\sqrt[10]{\arccos\left\{\lg\lg\lg\lg\left[\left\langle\frac{\mathbf{\&}_+^\times}{\alpha a_e}\right\rangle^{\Phi\Phi\Phi\Phi}\cdot\frac{\Phi^3 i_1}{\text{Spin}}\right]\right\}}=$$

$$=\mathbf{1.61803...}$$

Background energy fluctuation constant:

(182) $$\boldsymbol{Fluc}(tuation) = \left(\left(\frac{E_{ag}}{2}\right)^2\right)^2.$$

Astronomy parameters of the Solar system (84) according to James Jacobs' data are (183):

| | | |
|---|---|---|
| 2439 | **87,97046** | **0,387099** |
| 6051 | **224,69815** | **0,72336** |
| $\boldsymbol{R} = \mathbf{6378,14}\ kms$ | $\boldsymbol{OP} = \mathbf{365,25636053}\ Sol.days$ | $AU = \mathbf{1,000018}$ |
| 3393,4 | **686,9257** | **1,523638** |
| 71398 | **4332,23025** | **5,20248** |
| 60000 | **10800,4425** | **9,56329** |
| 25400 | **30953,4765** | **19,2937** |
| 24300 | **60839,6925** | **30,2743** |
| 1500 | **91305,195** | **39,6823** |

The Solar system topology (including Pluto) will tentatively be:

$$\odot_{+}^{\times} = (\text{Semi-major axes})_{+}^{\times} \cdot (\text{Orbital periods})_{+}^{\times};$$

(184) $$\odot_{+}^{\times} = \mathbf{52962026,5 \times 7,964169827 \times 10^{36}}$$

Ultrastring (or Pythagoras' monochord) topology (86):

$$\{\mathbf{X_M}, \mathbf{Y_F}\}_{+}^{\times} \to 1, \frac{15}{16}, \frac{8}{9}, \frac{5}{6}, \frac{4}{5}, \frac{3}{4}, \frac{5}{7}, \frac{2}{3}, \frac{5}{8}, \frac{3}{5}, \frac{9}{16}, \frac{8}{15}, \frac{1}{2};$$

(185) $$\{\mathbf{X_M}, \mathbf{Y_F}\}_{+}^{\times} = \left\{\begin{matrix}\mathbf{11585435253981,} \\ \mathbf{169654738516152}\end{matrix}\right\}_{+}^{\times} =$$

$$= 3,5623190924503326642740865131775e + 41.$$

The main space filling coefficients:

(186) Kepler's: **0,7408;**

(187) Rogers': **0,7797.**

There are certain mythological numbers said to be clues to the absolute knowledge such as (188-192)

Druids' Stonehenge code: **360/56;**
Anthique Greek: **310952;**
Ancient Mongolian: **1292049;**
**4328.5;**
**15804**, et cetera.

Previously we have had the integer, 3549965, that notably clarifies the entire situation with physics and mathematics. So, in terms of the Stonehenge code we shall have

(193) $$100.00...\frac{2\pi}{56} = e^{\Phi\pi i_1};$$

(194) $$\pi^{\chi\chi} = 56\cdot^{10.9999999...};$$

(195) $$100\left\{360 \div \lg\frac{\mathbf{3549965}^{\pi e}}{\text{Spin}}\right\} = e^{\Phi\pi i_1}.$$

What is immediately noticable

(196) $$\sqrt[\dim_\Phi \Phi^3 i_1]{e^{\Phi\pi i_1 \times \Phi^3 i_1}} = \mathbf{21047}\cdot$$

**The Secret History of the Mongols** dated back to 1240 AD contains an ultimate clue to the absolute geometry of space-time and matter as next

(197) $$\mathbf{4328.5}^{\Phi\pi e i_1} = \frac{10^{68.9999...}}{\Delta_1 \Theta_W \Theta_{STR.}}.$$

The hierarchy of the systems is in the descending order:

Cosmological reality;
Pure mathematics;
Theoretical physics.

As a consequence of Goedel's metatheorem, physics can only be unified by pure mathematics. The unified theory in turn may touch deeper hidden truths of reality existing beyond human will and above the current scientific paradigms. There can and must be truths that need neither proof, nor negation, says Goedel's proof. The universal mathematical machine is, in Newton's words, an intelligent Being. We cannot rule out the possibility that the Being encoded this geometry in the time rhythm of the Earth planet as below:

(198) $$\text{Orbital Period} = 365.256360...\ days;$$

(199) $$\Phi\cdot(\text{Sidereal Year})^{2\Phi\pi e i_1} = \frac{10^{88.9999...}}{(©@)^2};$$

(200) $$(\text{Sidereal Year})\left\{1+\frac{1}{\varepsilon}\right\}; \qquad \frac{1}{\varepsilon} = \frac{\pi^{\frac{\Phi^3\sqrt{i_1 i_2}}{\text{Spin}}}}{\&_+^\times}.$$

The logic of spontaneous harmonization process in the Solar system astronomical parameters resulted in this or any other perfection with most favourable consequences.

## Newton-Gauss Solution for Unification

> The changing of bodies into light, and light into bodies, is conformable to the course of Nature, which seems delighted with transmutations.
>
> **Sir Isaac Newton**

Cosmology did produce its observer who cannot but speculate about **the Creation Act**, which is, formally, the space-time and radiation-matter problem. The subject *a priori* obliged to perform this constructive act is the compass-and-straight edge algorithmic-minded minimax logic.

Therefore, the Newton-Gauss solution will, mathematically, be

(201) $$\sqrt[\sin\Delta_{Exprm.}]{\sqrt[\sin\Theta_W]{\Phi\cdot\perp\Delta_{Exprm.}\cdot i_1}} \equiv \{\mathbf{3,5,17,257,65537}\}_{+}^{\times}.$$

Such is the consequence of the two most fundamental equations

(202) $$x^N - 1 = 0;$$
$$y^2 - y - 1 = 0.$$

Accurately,

(203) $$\frac{\sqrt[\sin\Delta_{Exprm.}]{\sqrt[\sin\Theta_W]{\Phi\cdot\perp\Delta_{Exprm.}\cdot i_1}}}{1-\dfrac{10}{e^{\Phi\pi i_1}\cdot\pi\dfrac{\Phi^3\sqrt{i_1 i_2}}{\text{Spin}}}} = \{\mathbf{3,5,17,257,65537}\}_{+}^{\times}.$$

## Spontaneity

The only true logic is that of spontaneity when no logic is imposed from outside. At first was Nothing, but Something had arisen. The assumption is that the first was left on its own and the second evolved due to a spontaneous process subject to the innate minimax logic.

To us, Nothing is the mathematical point in the Euclidean sense. In pure, or ideal, theories there is no other way around as to think so. Thus, the Universe should be thought as the end result of some spontaneous mathematical transformations of a point. In fact, there is no transformations, but manifestions of the wealth of innate properties of the mathematical point.

## Bifurcation

Spontaneity is, if a bit roughly, equivalent to bifurcation. Consequetive bifurcations will be multifurcation.

Therefore, cosmology is reducible to the nonlinear multifurcation of a point, both geometrically and algebraically.

Bifurcation is a spontaneous and instantaneous event. If there is something at all, it bifurcates into its opposite or complimentarity. The result is duality.

For example, the wellknown constant Pi bifurcates in many ways, including a nontrivial

(204) $$\pi \leftrightarrow \frac{1}{\pi - 3}.$$

It seems crazy, but this kind of bifurcations of mathematical constants and operators serves the very mechanism for functioning of the physical Universe we observe. In the above we have, possibly, the prototype for potential energy. If so, then

(205) $$\sqrt[\pi-3]{\mathbf{U_E}} = a_e \cdot 10^{68.999...}.$$

The power 69 which can never be exact is often met in calculations because of symmetry and equilibrium reasons defined by

(206) $$\frac{2\pi}{5} + \frac{2\pi}{\cosh \cos^2 i} \neq 138.$$

As it is seen, the electron magnetic moment anomaly can be defined in many different ways. And it is natural, for in the universal system everything is determined by all the rest in combination. Any phenomenon emerges as the focal point of a row coinciding reasons that act, however, as one universal reason:

$$1+1+1+... \leftrightarrow 1\times1\times1\times...$$

## Universality

Universality is a stricter concept than it seems. Its rigour should have a self-test. Universality has to be founded upon some underlying mechanism. The test for and the mechanism of the universality of the system of mathematical harmony is encoded in natural numbers, e, g,

(207) $$e^{5\Phi\pi e i_1 \sqrt{2}} \{(\pi - 3)(\Delta_1 - 137)\} = \frac{3}{\mathbf{441683}} \cdot 10^{58}.$$

# Bifurcations
# of
# the Euclidean Geometry

> You must see the infinite, *i,e,* the universal, in your particular or it is only gossip.
> **Oliver Wendell Holmes**

It is the problem of choice. If concentrate on the Euclidean **postulate of parallels**, one will arrive at the non-Euclidean geometries of Lobachevsky and Riemann. Instead, we in turn may assume that the key problem in geometry is that of **temporal dimension**. So, approaches to the Euclidean system do bifurcate.

Since the school years we have Euclidean geometry and Ptolemean trigonometry, both being fully embraced by a circle. The circle is defined by its radius. Theoretical sciences consider usually the unit circle where the radius is denoted by the unity, 1. Consequences of this operation are well known, but quite not satisfactory. It became clear that without time dimension constructed in geometry explicitly nothing to do in physics where rules time. The Greeks themselves didn't miss to consider this problem, of course. But, they thought, unfortunately, that the time axis cannot and need not to be constructed in geometry because of the illusory nature of time. It should be noticed that in the special relativity we have, in fact, the only two-dimensional space-time subject to the Lorenz transformations. The same can be said about gravitation. This kind of speculation leads to the idea that the time axis of geometry can and must be drawn on a plane. But how?

The problem lies in bifurcations. We have to have a circle other than the unit circle. Characteristics of circles depend on their radius. The unit circle with the radius, 1, ends up with the Euclid-Ptolemean system:

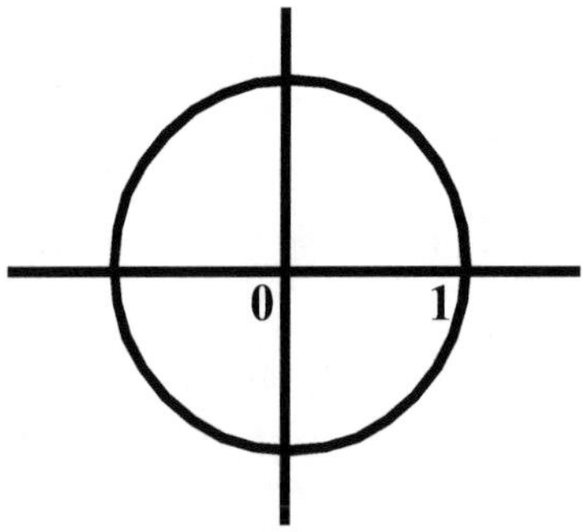

Now we denote the raduis as **Φ** with the consequence that we smoothly shift to the post-Euclidean geometry made with the temporal dimension in accordance with the real physical World (**Picture G**):

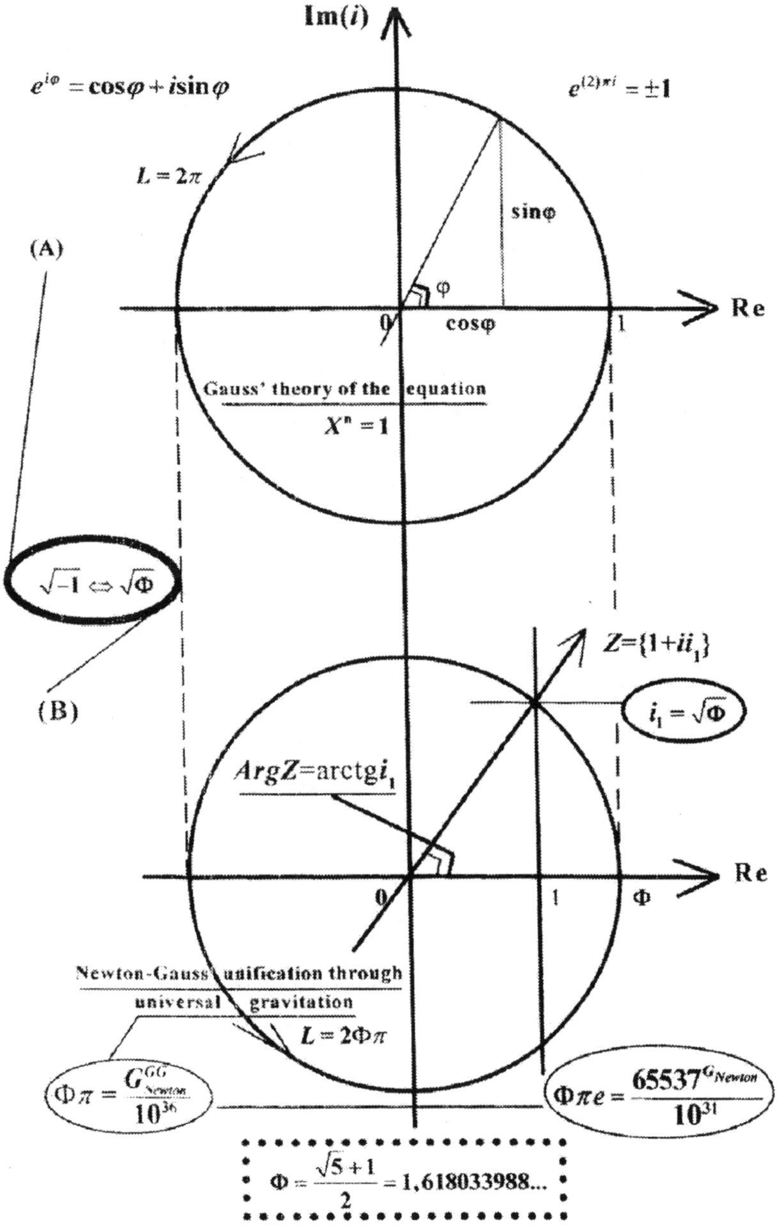


As it is seen, this picture alone suffices to unify Newton's physics and Gauss' pure mathematics, as it should be so owing to the minimax logic of things. The reason is simply that any given circle is approximated by the regular polygons constructible algorithmically with the compass and straightedge. Return back to the formula (71).

Gravity is universal. Now we see the fundamental reasons of why it is so. Newton's constant is a fundamental mathematical fact.

## Matter and Radiation

Do not scorn particulars;
they are universals made easy.
**Henry S. Haskins**

We write a unified radiation-matter configuration

(208) $$(\Phi \perp i_1)(\Phi\Delta_{Exprm.}i_1) = \mathbf{RM}\,.$$

A more complete way of writing things will certainly be

(209) $$\mathbf{RM}\cdot\left\{\mathbf{D\{2\}}^{\times}_{+}\cdot\mathbf{D\{3\}}^{\times}_{+}\cdot\mathbf{D\{4\}}^{\times}_{+}\right\}\cdot\mathbf{InG}^{\times}_{+} = 2\pi Rad\cdot 10^{93}\,,$$

where geometry is decomposed into topological phenomena.

If the other way around, then by definition

(210) $$\mathbf{RM}^{\Phi\pi e} \equiv \frac{1}{e^{\Phi\pi i_1}}\,.$$

And naturally,

(211) $$(\Phi\Delta_{Exprm.}i_1)^{\Phi\pi e} = \frac{\mathbf{64816}}{9}\cdot 10^{30}\,;$$

(212) $$\mathbf{64816}^{\pi\Phi^3 i_1} = \mathbf{D\{4\}}^{\times}_{+}\cdot 10^{36}\,.$$

The loop of the logic of universality is always closed as it is the case in

(213) $$\sqrt[3+1]{\sqrt[\dim_{\Phi}\Phi^3 i_1]{\mathbf{64816}^{\pi\frac{\Phi^3 i_1}{\text{Spin}}}}} \equiv (\Phi \perp i_1)(\Phi\Delta_{Exprm.}i_1)\cdot$$

Accurately,

(214) $$\ldots = 100(\Phi \perp i_1)(\Phi\Delta_{Exprm.}i_1)\left\{1+\sqrt[\Phi\Phi]{\frac{1}{e^{\Phi}\cdot 10^{10}}}\right\}.$$

It should be kept in mind that beauty and symmetry of the unifying mathematics do, of course, say for themselves. Yet, for the physical world are incredibly more vital not symmetries, but their violations. The universal machine is doomed to work eternally simply because it cannot avoid calculations of symmetry violations which, fortunately, never cease.

If rigid symmetries were possible, the machine would have stopped to work a cosmological time ago.

Therefore, **the existence of the physical Universe owes to the inprecision of the system of mathematical harmony.** This can be called the gauge, or compensation symmetry, principle.

## Elements of the Universal Calculus

> Philosophy asks the simple question:
> What is it all about?
> **Alfred North Whitehead.**

For brevity's sake just an example. In the modern physics we have a certain definite set of fundamental constants and all of them can be derived in one go as below by definition:

(215) $$10\frac{Ghm_e e^{\pm}c\cdot\dim(Ghm_e e^{\pm}c)}{\alpha a_e}\equiv\frac{\mathbf{InG}_{+}^{\times}}{e^{\Phi\pi i_1}}\,.$$

The experimental values of constants we rely upon throughout this work do structly satisfy the previous equation if the left section is corrected by

$$1+\varepsilon\cdot$$

Therefore,

(216) $$\frac{1}{\varepsilon}\cdot\pi\frac{\Phi^3\sqrt{i_1 i_2}}{\mathrm{Spin}}\cdot\frac{\mathbf{D\{2\}}_{+}^{\times}\mathbf{D\{3\}}_{+}^{\times}\mathbf{D\{4\}}_{+}^{\times}}{\mathbf{InG}_{+}^{\times}}\cdot\mathbf{U_E}=\exists,$$

which is a typical picture of the self-perturbation of the entire geometry.

The moral here is that the experimental values of physical constants can no more be improved, for no precision experiments can eliminate self-perturbative effects of geometry itself.

Besides, it is not exactly that in the above we derive the main set of constants: in the system of universal harmony there are incredibly many ways to do the same. Introducing a shorthand notation **A** for the quintuplet of constants, approximate and analyse, for example,

(217-218) $$\sqrt[\Phi]{\exp e^{e}\cdot 10^{9}}\equiv\mathbf{A}\,;\qquad \sqrt[\cos i]{\frac{10^{16}}{\Phi\pi e}}\equiv\mathbf{A}\,;$$

(219) $$\sqrt[\cos 2\Theta_{STR.}]{\frac{\mathbf{A}}{\cos 2\Theta_{STR.}}}\equiv\frac{10^{71}}{\mathbf{D\{4\}}_{+}^{\times}}\,.$$

Practically exact a topology is

(220) $$\mathbf{A}\cdot\frac{\mathbf{D\{4\}}_{+}^{\times}}{\mathrm{Sing}\cdot\dim_{\Phi}\Phi^3 i_1\cdot(3+1)}=\frac{10^{57}}{\mathbf{22658}}\,.$$

Physics is such, not otherwise, because constants owe to all the overlapping reasons of mathematical harmony. Since all roads lead to Rome, the unifield equations one can reasonably imagine and experiment with are always **fool-proof**. So, enhance the previous composition **A** on the account of nuclear force parameters and get the unified physics at one go as next:

$$(221):\quad \sqrt[\Phi]{\left\langle \mathbf{A}\frac{\Theta_W \Theta_{STR.}}{\sin\Delta_{Exprm.}\sin\Theta_W\cos\Theta_W\cos 2\Theta_{STR.}}\right\rangle^{\pi e}} = \\ = \mathbf{321608}\cdot 10^{66}.$$

Is it probable that, indeed,

$$(222)\qquad \exp\frac{\mathbf{Homo}_{\mathbf{542891047}}\cdot e^{2\Phi\pi i_1}\cdot e^{5\Phi\pi e i_1\sqrt{2}}}{10^{69}} = \frac{2}{\Phi}\ ?$$

If check flicking through the literature, the Fi constant occurs only once in quantum theory. If believe Max Born, the logos on the right of the above is the coefficient to the electron radius (See: **Atomic physics** by Max Born). Possibly, Born himself didn't notice the occurence. Or he didn't take the risk of being laught at for saying of some importance of $\Phi$ in theoretical physics.

Whatsoever, the Darwinian species Homo Sapiens Sapiens came to the scene as a result of physical evolution during the cosmological time span till now. Therefore, we try and get a logos

$$((223)\qquad 10\frac{\mathbf{Homo}_{\mathbf{542891047}}}{\mathbf{A}=\left\{\dfrac{Ghm_e e^{\pm}c\cdot\dim(Ghm_e e^{\pm}c)}{\alpha a_e}\right\}} = \frac{4}{\pi} = i_2 \ .$$

We by no means persist on that the integer under discussion is the cosmological code for man. Nonetheless, the following is the typical mechanism of doing things in this geometry:

$$(224)\qquad \frac{\mathbf{542891047}}{\sin\Theta_W\cos 2\Theta_{STR.}} = \left\{\cos\frac{2\pi}{5} = \frac{1}{2\Phi}\right\}\cdot 10^{10}$$

The matter is that the absolute geometry has only and only one figure for all purposes. Such is the minimax logic. This unique figure is of the symmetry of the fifth order. And it has handedness, that is, chirality implying that right and left are not the same in the real geometry of Nature.

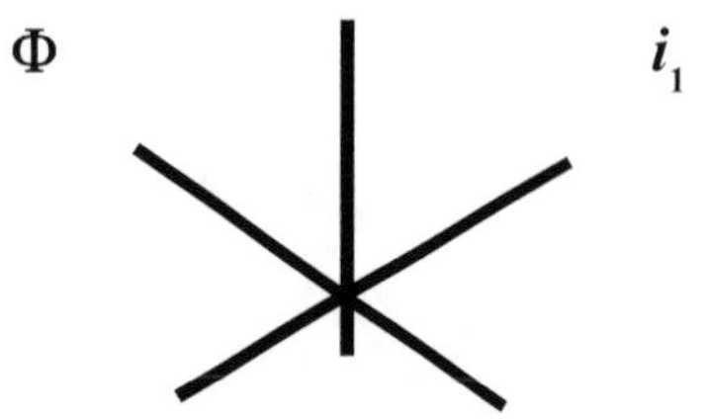


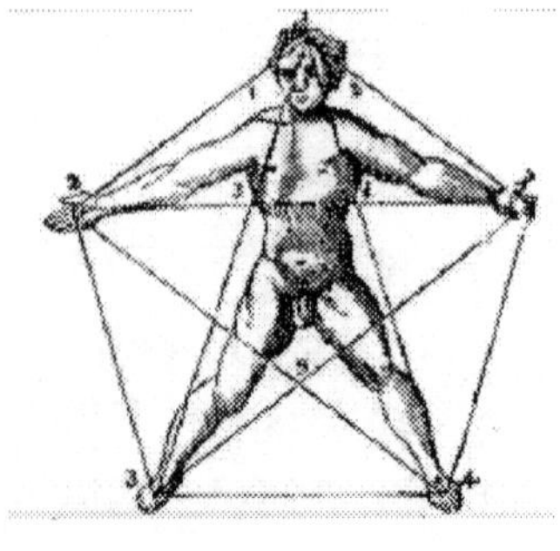

## Jiordano Bruno's Unification Theorem

Jiordano Bruno appears to have unified Pythagoras' (arithmetization) and Plato's (geometrization) concepts and formulated, probably, the most genial theorem everknown:

**"The order of a unique figure and**
**the harmony of a unique number**
**give rise to all things."**

(We quote Bruno's theorem according to **Alchemy & Mysticism** by Alexander Roob who refers to: Jiordano Bruno, ***About the Mona***, 1591.)

The unique figure of absolute geometry results from the algorithmic multifurcation of mathematical point as follows below (**Picture I**):

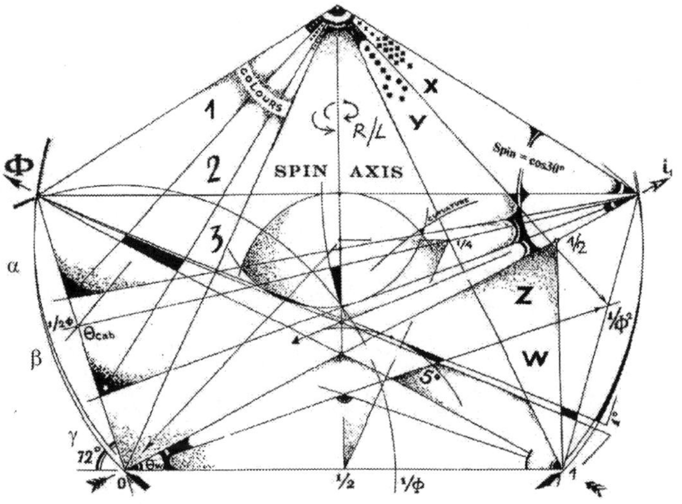


The figure is the nonlocal point-singularity that serves the unified model for all the fundamental fermions. Algebraically,

(225) $$\sqrt[\Phi\pi i_1]{\left\{\left\{\pi\frac{\Phi^3\sqrt{i_1 i_2}}{\text{Spin}}\right\}^2\cdot\dim_\Phi \Phi^3 i_1\cdot(3+1)\cdot\frac{\mathbf{D\{2\}}^\times_+\cdot\mathbf{D\{3\}}^\times_+\cdot\mathbf{D\{4\}}^\times_+}{\mathbf{InG}^\times_+}\right\}^e}=\mathbf{2866}\cdot 10^{25}.$$

What is strange, something here is squared. Yet, consistently,

(226) $$\sqrt[\pi\frac{\Phi^3 i_1}{\text{Spin}}]{\mathbf{67151}^{\dim_\Phi \Phi^3 i_1\cdot(3+1)}}=\mathbf{2865.999999...}$$

## Mathematics as a Science, Not Sports

The modern mathematics is any doubt a great achievement. But, it stagnates; the situation currently looks fairly rotten. Maths became something like intellectual sports at best, or even just a scholastic game in diehard formalism of axiomatic sort, at worst. Say, Fermat's Last theorem is merely a sports problem which is of hardly use even for mathematics itself.

Instead, we pursue the idea that mathematics, in general, and its pure branch, in particular, does pertain to the natural sciences, as physics does so. At the same time, more distanced between them sciences as pure maths and theoretical physics one cannot imagine at the moment. This ackward situation now undergoes the most radical revolution, notably, that physics and pure maths become unified through natural numbers alone.

I like sports. Nobody now believes, but it is really true that in my time as boxer I was trained by the famous Mr Jackson, twice US champion in the middle weight category. You never know who was who and where in the turbulent 20th century. I like sports, but I don't like maths as sports. Mathematics is a natural science.

## The Fundamental Theorem of Pure Mathematics

Karl Gauss unified higher arithmetic (number theory), algebra and geometry into what is called pure mathematics. The latter, however, still remains without its fundamental theorem despite that arithmetic and algebra, or even analysis, have their own.

Therefore, the following logos fills the gap:

(227) $$\frac{1}{10©@} \equiv \frac{\cos^2 i}{\Phi^3 i_1}.$$

Symmetries are infinitesimally broken as

(228) $$1.000000... = 10©@ \cdot \frac{\cos^2 i}{\Phi^3 i_1}.$$

Given a self-growth process such that

$$\left\{\frac{\exists}{©@} = 25.23671000961...\right\}^{\pi \frac{\Phi^3 \sqrt{i_1 i_2}}{\text{Spin}}} \equiv \frac{1}{\exp e^e},$$

we find

(229) $$1+\frac{\exists}{©@}\cdot\frac{1}{\mathbf{34094965}}=10©@\cdot\frac{\cos^2 i}{\Phi^3 i_1}.$$

The integer, 34094965, seems senseless, but the very point is that it is not so, for

(230) $$\frac{\mathbf{34094965}}{\alpha a_{Electron}}=\mathbf{34094965}\cdot\Delta_{Exprm.}\aleph_1=\mathbf{4029}\cdot 10^9.$$

By the way, simply

(231) $$\{\Delta_{Exprm.},\aleph_1\}_+^\times=\sqrt[10]{\mathbf{5276}\cdot 10^{77}},$$

and

(232) $$\sqrt[\Phi]{(\mathbf{5276}\times\sqrt[\Phi\pi e+1]{10^{90}})^{\pi e}}=e^{\Phi}\cdot 10^{51}.$$

In connexion with the fundamental theorem we would have

(233) $$\left\{\pi\frac{\Phi^3\sqrt{i_1 i_2}}{\text{Spin}}\cdot\frac{\exists}{©@}\right\}^{\pi e}=\frac{10^{37}}{\mathbf{16299}};$$

(234) $$\left\{\pi\frac{\Phi^3\sqrt{i_1 i_2}}{\text{Spin}}\right\}^{\frac{\exists}{©@}}\cdot\left\{\frac{\exists}{©@}\right\}^{\pi\frac{\Phi^3\sqrt{i_1 i_2}}{\text{Spin}}}=\sqrt[\Phi]{\frac{2}{\Phi}\cdot 10^{97}}.$$

Max Born's coefficient to the electron radius should have, indeed, once occasionally occured in quantum mechanics, for

(235) $$\sqrt[\sin\Delta_{Exprm.}]{\left\langle\frac{2}{\Phi}\right\rangle^{\Delta_{Exprm.}}}=\sqrt[\Phi\pi]{\mathbf{118563}\cdot 10^{89}}=\sqrt[\Phi\pi]{\Delta_1\aleph_1\cdot 10^{88.9999...}}.$$

## Why Human Chromosome Number is 46 ?

In this theory wording is unnecessary. So, *Nullius in Verba*:

(236) $$\sqrt[5]{(\Phi\pi e i_1)^{46}}=\mathbf{284}\cdot 10^9;$$

(237) $$\left[\{2,8,4\}_+^\times\cdot 284\right]^{\pi}=\mathbf{9600106}\cdot 10^{10};$$

(238) $$\sqrt[e]{\left[\{2,8,4\}_+^\times\cdot 284\right]^{2\Phi\pi i_1}}=\frac{\mathbf{1562}}{\mathbf{3}}\cdot 10^{23};$$

(239) $(\Phi\pi e i_1)^{46}\cdot 46^{\Phi\pi e i_1} = \mathbf{176118^2}\cdot 10^{76}$ ;

(240) $(\Phi\pi e)^{76} = \mathbf{777936^3}\cdot 10^{69}$ ;

(241) $$\sqrt[\pi\frac{\Phi^3 i_1}{\text{Spin}}]{\left\{(\Phi\pi e i_1)^{46}\cdot 46^{\Phi\pi e i_1}\right\}^{\dim_\Phi \Phi^3 i_1\cdot(3+1)}}\cdot\frac{\mathbf{3383}}{3} = 10^{65};$$

(242) $$\mathbf{3383}^{XX} = \frac{10^{95}}{\mathbf{31602}},\ \textit{et cetera, ad infinitum.}$$

Evolution of matter tends to the mathematical system of harmony and the process so far ends up with human genetics.

## The Constant Aleph-One

Don't be 'consistent,' but be simply true.
**Oliver Wendell Holmes**

The post-Dirac effect in the qunatum-relativistic physics can be computed in many ways affordable in this geometry, e, g,

(243) $$\sqrt[\sin\Delta_{Exprm.}\sin\Theta_W\cos\Theta_W\cos 2\Theta_{STR.}]{\aleph_1} = \frac{10^{30}}{\mathbf{386}}.$$

The electron magnetic moment anomaly is a universal constant of the new fundamental theory, as it should be so.

If translate the modern quantum electrodynamics method of computing the anomaly into the continuum theory, then

(244) $$\frac{\aleph_1}{2\pi\cdot\Delta_{Exprm.}} = 1+\frac{e^e}{10000}\left\{1+\frac{e^e}{10000}\right\}\frac{1}{10000}\cdots$$

The technique of Feynman diagrams in QED and the notion of virtual particles are both not quite false from the point of view of permanent bifurcation-oscillations of geometry. The only remark is that Nature does not calculate things that way which takes colossal amount of labour in man/hour and supercomputer/hour terms. God is a wholesale trader exploiting the advantages of the universal calculus in doing things instantly.

For example,

(245) $$\left\{Ghm_e e^{\pm}c\cdot\dim(Ghm_e e^{\pm}c)\right\}^{2\Phi\pi e i_1} = \mathbf{158712}\cdot 10^{155},$$

whence experimental physics should comply with

(246) $$Ghm_e e^{\pm}c = 1934.62600882...$$

## Wien's Displacement Law

The black body radiation problem remains a classical topic in physics. In this respect it mighy be interesting to play, e., g. with the Wien displacement law usually expressed by the constant

$$\lambda_{\max} T .$$

Then, the following representation, or rather, explaination can be guessed in some way or other:

(247)
$$\frac{\mathbf{U}_{\mathrm{E}}}{\lambda_{\max} T = \mathbf{0.289776743...}} = \mathbf{78} \cdot 10^{8} .$$

Also,

(248)
$$\pi \frac{\Phi^{3}\sqrt{i_1 i_2}}{\mathrm{Spin}} \cdot \frac{\mathbf{D\{4\}}_{+}^{\times} \cdot \mathbf{InG}_{+}^{\times} \cdot \mathbf{U}_{\mathrm{E}}}{\lambda_{\max} T} = \Delta_1 \cdot \frac{\mathbf{65537}}{\Phi^{9}} \cdot 10^{65} (1-\varepsilon)$$

(249)
$$\frac{1}{\varepsilon} = \sqrt[\Phi \pi e i_1]{1000 \mathbf{D\{4\}}_{+}^{\times}} .$$

Consequently, important is a bifurcation

(250)
$$\left\{ \frac{1}{\alpha a_e} = \Delta_{Exprm.} \aleph_1 \right\} \Leftrightarrow \left\{ \Delta_1 \cdot \frac{\mathbf{65537}}{\Phi^{9}} \right\}$$

This is justified by (251)

$$\pi \frac{\Phi^{3}\sqrt{i_1 i_2}}{\mathrm{Spin}} \cdot \dim_{\Phi} \Phi^{3} i_1 \cdot (3+1) \times \Delta_1 \cdot \frac{\mathbf{65537}}{\Phi^{9}} = \sqrt[\pi e]{\mathbf{13958} \cdot 10^{60}} .$$

The best way to approximate the two values concerned is

(252)
$$1 = \frac{\Delta_1 \cdot \frac{\mathbf{65537}}{\Phi^{9}}}{\Delta_{Exprm.} \aleph_1} - \left\{ \varepsilon = \frac{1}{\sqrt[\chi\chi]{\Phi^{3} i_1 \cdot 10^{63}}} \right\} .$$

Besides,

(253)
$$\Delta_1 \cdot \mathbf{65537} = \frac{\mathbf{27035539}}{3} .$$

The above example with Wien's law is chosen just to say to the reader that he/she may pick up whatever in physics and rethink from the unified-theory theoretical point of view.

## Unified Physics at Once and forever ?

The very notion of the unified physics does imply that it is an imperative to write a single equation that could describe the current state of physical cosmology as a whole.

For this purpose we have such basic components as

(254) $$\mathrm{A} = \left\{ \pi \frac{\Phi^3 \sqrt{i_1 i_2}}{\mathrm{Spin}} \right\}^2 \dim_\Phi \Phi^3 i_1 \cdot (3+1) \cdot \frac{\mathbf{D\{2\}}^\times_+ \mathbf{D\{3\}}^\times_+ \mathbf{D\{4\}}^\times_+}{\mathbf{InG}^\times_+};$$

$$\mathrm{B} = N \cdot Ghm_e e^{\pm} c \cdot \dim(Ghm_e e^{\pm} c);$$

$$\Gamma = \frac{\Delta_{Exprm.}}{\sin \Delta_{Exprm.}} \aleph_1;$$

$$\Lambda = \frac{\Theta_{\text{Electroweak}} \Theta_{\text{Nuclear strong force}}}{\sin \Theta_W \cos \Theta_W \cos 2\Theta_{STR}};$$

(255) $$\mathrm{M} = 2 \times 12_{\text{quarks and leptons}} \cdot$$

Then, the total topology will be

(256) $$\mathrm{AB\Gamma\Lambda M} = \mathbf{7.653884963} \cdot 10^{85},$$

which would look quite not inconsistent, for

(257) $$\sqrt[\Phi]{\mathrm{AB\Gamma\Lambda M}} = \mathbf{12} \cdot 10^{52};$$

$$\sqrt[\pi e]{(\mathrm{AB\Gamma\Lambda M})^{\Phi i_1}} = \mathbf{5} \cdot 10^{20}.$$

Assuming that the Solar system intelligent being is

(258) $$\left\{ \odot^\times_+ \cdot \frac{\text{🕸}}{\text{😐}} \right\},$$

we find that

(259) $$\left\{ \odot^\times_+ \cdot \frac{\text{🕸}}{\text{😐}} \right\} \times \left\{ \odot^\times_+ \cdot \frac{\text{🕸}}{\text{😐}} \right\} = \pi\Pi_{Cosm.} \cdot 10^{140}.$$

The obviously fundamental entity

(260) $$\pi\Pi_{Cosm.}$$

is met in calculations more often than not.

## Superstructure of Space-Time & Matter

Carefully investigating the **Superstructure of space-time and matter** one can pick up any of indeces made visible below:

(261)
$$ {}^{\Phi i_1}_{i_1\Phi} \boldsymbol{SS}^{\mathbf{1,2,3}\ \uparrow\downarrow\ \boldsymbol{X,Y}}_{\pm\boldsymbol{\alpha,\beta,\gamma}\ \rightleftarrows\ \boldsymbol{Z,W}} $$

Thus, we may write fundamental fermion particles as

(262)

| | | |
|---|---|---|
| $\nu_e^X$ | $\nu_\mu^Y$ | $\nu_\tau^Z$ |
| $e_{\alpha\beta\gamma}^{XZ}$ | $\mu_{\alpha\beta\gamma}^{YW}$ | $\tau_{\alpha\beta\gamma}^{XYZW}$ |
| $u_{\beta\gamma}^1$ | $c_{\alpha\gamma}^2$ | $t_{\alpha\beta}^3$ |
| $d_\alpha^{2,3}$ | $s_\beta^{1,3}$ | $b_\gamma^{1,2}$ |

The origin of the superstructure lies in the quantum leap process

(263)
$$ \left\{ \begin{array}{l} \pi \dfrac{\Phi^3\sqrt{i_1 i_2}}{\text{Spin}} \cdot \dim_\Phi \Phi^3 i_1(3+1)\cdot \\ \cdot \dfrac{\mathbf{D\{2\}}_+^\times \cdot \mathbf{D\{3\}}_+^\times \cdot \mathbf{D\{4\}}_+^\times}{\mathbf{InG}_+^\times} \cdot \mathbf{U_E} \cdot E \end{array} \right\} = \frac{10^{80}}{(©@)^2}\,. $$

Then the $X$-constant fills geometry with the physics content (264):

$$ X \left\{ \begin{array}{l} \pi \dfrac{\Phi^3\sqrt{i_1 i_2}}{\text{Spin}} \dim_\Phi \Phi^3 i_1(3+1)\cdot \\ \cdot \dfrac{\mathbf{D\{2\}}_+^\times \mathbf{D\{3\}}_+^\times \mathbf{D\{4\}}_+^\times}{\mathbf{InG}_+^\times} \mathbf{U_E} \cdot E \end{array} \right\} \left\{ \frac{Ghm_e e^\pm c \cdot \dim(\ldots)}{\alpha a_e} \right\} = $$

$$ = \pi^2 \cdot 10^{95.9999\ldots}. $$

We have to have necessarily an integer, too (265),

$$\sqrt[\pi e]{\left\{\begin{array}{l}\pi\dfrac{\Phi^3\sqrt{i_1 i_2}}{\text{Spin}}\dim_\Phi \Phi^3 i_1(3+1)\cdot\\ \cdot\dfrac{\mathbf{D\{2\}_+^\times D\{3\}_+^\times D\{4\}_+^\times}}{\mathbf{InG_+^\times}}\cdot\mathbf{U_E}\cdot E\end{array}\right\}^{\left\{\frac{Ghm_e e^{\pm} c\cdot\dim(...)}{\alpha a_e}\right\}}} =$$

$$=\frac{10^{16}}{\mathbf{225554}}.$$

To spare space, in this introductory book we omit detailization of the combinatorics offered by the Superstructure, and leave this job rather to younger students for exercizes. Much more important will be general analysis of physics by the method of synthesis.

The Platonic regular topology provides

(266) $$\left\langle\mathbf{\{3,5,17,257,65537\}_+^\times}\cdot\Phi\right\rangle^\Phi=\mathbf{5258498}\cdot 10^{17};$$

(267) $$\sqrt[\dim_\Phi \Phi^3 i_1\cdot(3+1)]{\mathbf{D\{3\}_+^\times}^{\pi\frac{\Phi^3 i_1}{\text{Spin}}}}=\frac{3}{\mathbf{1118}}\cdot 10^{27};$$

(268) $$\frac{\mathbf{D\{2\}_+^\times\cdot D\{3\}_+^\times\cdot D\{4\}_+^\times}}{e^{3\Phi\pi i_1}}=\mathbf{945937}\cdot 10^{63};$$

(269) $$\sqrt[e]{\left\{\pi\frac{\Phi^3 i_1}{\text{Spin}}\dim_\Phi \Phi^3 i_1\cdot(3+1)\cdot\mathbf{D\{4\}_+^\times}\right\}^\pi}=\nabla\cdot 10^{53};$$

(270) $$\pi\frac{\Phi^3\sqrt{i_1 i_2}}{\text{Spin}}\dim_\Phi \Phi^3 i_1\cdot(3+1)\cdot\mathbf{D\{4\}_+^\times D\{3\}_+^\times D\{2\}_+^\times}=$$

$$=\frac{\mathbf{619265}}{9}\cdot 10^{75};$$

(271) $$\mathbf{619265}^{2\Phi\pi i_1}=\frac{9}{\mathbf{113357}}\cdot 10^{79}.$$

The moral is here and else where is that God considers, indeed, natural numbers alone, while we invent fractions for our purposes for understanding and computing.

## Gravi-electromagnetism

Execution is the chariot of genius.
**Wiiliam Blake**

Gravi-electromagnetism is approachable in many ways, including

(272) $$\{[\mathrm{Sing}\cdot\dim_{\Phi}\Phi^{3}i_{1}\cdot(3+1)]\Delta_{\mathit{Exprm.}}\}^{G\dim G}=\frac{10^{63}}{\mathbf{3112879}}.$$

And it is quite not hard to guess that

(273) $$\mathbf{3112879}\cdot\frac{\mathbf{D\{4\}_{+}^{\times}}}{\mathbf{InG_{+}^{\times}}}\cdot\mathbf{U_{E}}=\frac{10^{50}}{\Delta_{1}}.$$

Just an exersize in computing fundamental constants:

(274) $$\Delta_{1}^{2\Phi\pi e i_{1}}=\frac{10^{82}}{\mathbf{6800216}},$$

where

$$\pi=3.141\ 592\ 653\ 58...$$

## Human Genome

Etiquette requires us to admire
the human race.
**Mark Twain**

What is, indeed, admirable,

(275) $$\mathbf{\{Genes\}_{+}^{\times}}=\mathbf{1.657760095}\cdot10^{75};$$

(276) $$\mathbf{\{Genes\}_{+}^{\times}}\cdot\mathbf{D\{4\}_{+}^{\times}}=\frac{\mathbf{140935}}{3}\cdot10^{116};$$

(277) $$\mathbf{\{Genes\}_{+}^{\times}}\cdot\frac{\mathbf{D\{4\}_{+}^{\times}}}{\mathbf{Cr_{+}^{\times}}}=\mathbf{6}\cdot10^{101};$$

(278) $$\mathbf{\{Genes\}_{+}^{\times}}\cdot\frac{\mathbf{D\{4\}_{+}^{\times}}\cdot\mathbf{Cr_{+}^{\times}}}{\mathbf{Ur_{+}^{\times}}}=(©@)^{2}\cdot10^{121.0000...}.$$

If physically, then

$$(279)\qquad \{\mathbf{Genes}\}_{+}^{\times}\frac{Ghm_e e^{\pm}c\cdot\dim(...)\cdot\Delta_{Exprm.}\aleph_1}{\cos 2\Theta_{STR.}}=180\cdot 10^{83},$$

leading to the pentasymmetry

$$\mathbf{180}\cdot 10^{83}=\sqrt[\cos 2\Theta_{STR.}]{\cos 72\cdot 10^{34}}\ ;$$

$$\sqrt[\Phi]{\{\mathbf{Genes}\}_{+}^{\times}}=\tan 72\cdot 10^{46}.$$

The reader may tie knots with a lengthy enough strip of paper so that to see how the double helix structure evolves.

Finally,

$$(280)\qquad \frac{10^{101}}{\{\mathbf{Genes}\}_{+}^{\times}\cdot\{\text{🕸}\cdot\text{😐}\}}-1=\Phi^3 i_1$$

## Newton-Gauss-Planck Unification

Nature is visible thought.
**Heinrich Heine**

First, immediately,

$$(281)\qquad \sqrt[e]{\left\{\frac{G_{Newton}}{\Pi_{Cosm.}}\cdot\{\mathbf{3,5,17,257,65537}\}_{+}^{\times}\right\}^{2\Phi\pi i_1}}=\frac{\mathbf{2638453}}{3}\cdot 10^{68};$$

$$(282)\qquad \mathbf{2638453}^{\pi=3,14159265...}=\frac{\mathbf{44708}}{3}\cdot 10^{16};$$

$$(283)\qquad \mathbf{2638453}^{\mathbf{6,6730000000...}\Pi}=\mathbf{1401511}\cdot 10^{17};$$

$$(284)\qquad \mathbf{1401511}^{\Phi\Phi}=\frac{2}{\Phi}\cdot 10^{16},$$

and the sequence shifs to Born's coefficient in QM.

Secondly (285),

$$\left\{\left\{Gh\cdot\Phi^3 i_1\right\}\{\mathbf{3,5,17,257,65537}\}_{+}^{\times}\right\}\cdot e^{\Phi\pi i_1}=\mathbf{432977}\cdot 10^{14};$$

$$(286)\qquad \mathbf{432977}^{G\Pi_{Cosm.}}=\frac{10^{20}}{\ln\Phi}.$$

## Pythagoras' Monochord-Ultrastring

Learn get used to it. Eels get used to skinning.
**Winston S. Churchill**

It is foreseeable that

(287) $$\{\mathbf{X_M}, \mathbf{Y_F}\}_+^{\times\Phi} \cdot \Phi = \mathbf{828371} \cdot 10^{62}$$

The ultrastring would sound rather pentatonic:

(288) $$\{\mathbf{X_M}, \mathbf{Y_F}\}_+^{\times} \cdot e^{5\Phi\pi i_1} = \mathbf{39114} \cdot 10^{51};$$

(289) $$\mathbf{39114}^{\chi \cdot G\Pi} = 2\Theta_{STR.} \cdot 10^{82}.$$

The light cannot but be the ultrastring-monochord itself:

(289) $$\sqrt[e]{\left\{\{\mathbf{X_M}, \mathbf{Y_F}\}_+^{\times} \cdot \{\Phi \cdot 90 \cdot i_1\} \cdot c\right\}^{\pi}} = \frac{10^{66}}{1 \cdot 2 \cdot 3 \cdot 4 \cdot 5 \cdot 6 \cdot 7 \cdot 8 \cdot 9}.$$

Now that the Pythagoras-Plato logos

(290) $$4\pi \frac{\mathbf{D}\{\mathbf{4}\}_+^{\times}}{\{\mathbf{X_M}, \mathbf{Y_F}\}_+^{\times}} = 100000.$$

## Kinematics of Energy

The Delta angle defines kinematics of energy

(291) $$\sqrt[\sin\Delta_{Exprm.}]{\sqrt[\sin\Delta_{Exprm.}]{\mathbf{U_E}}} \equiv \Delta_1;$$

(292) $$\sqrt[\sin\Delta_{Exprm.}]{\sqrt[\sin\Delta_{Exprm.}]{\sqrt[\sin\Delta_{Exprm.}]{E}}} \equiv \cos 72;$$

$$\sqrt[\sin\Delta_{Exprm.}]{\sqrt[\sin\Delta_{Exprm.}]{\sqrt[\sin\Delta_{Exprm.}]{\widehat{E}\breve{E}}}} \equiv 6 \cdot$$

Leaving intermediary results, we obtain

(293) $$\frac{\widehat{E}\breve{E} \cdot \mathbf{U_E} \cdot E}{\sin\Theta_W \cos\Theta_W \sin\Delta_{Exprm.} \cos 2\Theta_{STR.}} \equiv \Pi_{Cosm.}$$

Reminding a simple equality that

(294) $$\cos\frac{2\pi}{5} = \frac{1}{2\Phi}$$

and also

(295) $$\sqrt{\frac{\pi e}{2}} = \text{Ramanujan Series},$$

we in turn discover an asymptotical equality by definition

(296) $$\sqrt{\frac{\pi e}{2}} \equiv \sqrt[\Phi]{2\Phi}.$$

If accurately,

(297) $$\sqrt{\frac{\pi e}{2}} = \sqrt[\Phi]{2\Phi\left(1-\frac{\cos 2\Theta_{STR.}}{100000}\right)}.$$

Despite the colossal mix of constants, the following is accurate:

(298) $$\sqrt[5]{e^{5\Phi\pi e i_1\sqrt{2}}\cdot \widehat{E}\breve{E}} = \mathbf{1889404}\cdot 10^{9},$$

proving that the energy-entropy configuration does exist.

## Foundations of Gravity

First, some elementary tricks that however conceals much subtleties of the problem of universal gravity:

(299) $$65537^{G} \equiv \Phi\pi e\cdot 10^{31};$$

(300) $$G^{GG} \equiv \Phi\pi\cdot 10^{36};$$

(301) $$65537\cdot G_{6.673000000...} = \sqrt[\pi]{\mathbf{5261626}\cdot 10^{11}};$$

(302) $$\sqrt[©@]{\{\mathbf{6,6,7,3}\}_{+}^{\times}\cdot 6673} \equiv \{\mathbf{X_M},\mathbf{Y_F}\}_{+}^{\times};$$

(303) $$X\left\{\{\mathbf{6,6,7,3}\}_{+}^{\times}\cdot 6673\right\}\cdot 65537 \equiv \frac{1}{a_e},$$

*et cetera ad infinitum.*

## Stability of the Solar System

The stability of the given Solar system configuration is a fundamental cosmological and anthropological problem. So,

(304) $$\sqrt[(3+1)\dim_\Phi \Phi^3 i_1]{\left\{\odot_+^\times \cdot \frac{Ghm_e e^\pm c \cdot \dim(\ldots)}{\alpha a_e}\right\}^{\pi\frac{\Phi^3\sqrt{i_1 i_2}}{\mathrm{Spin}}}} \equiv \frac{1}{\&_+^\times};$$

(305) $$\odot_+^\times \cdot \left\{\pi \frac{\Phi^3\sqrt{i_1 i_2}}{\mathrm{Spin}} \dim_\Phi \Phi^3 i_1 (3+1) \cdot \mathbf{D\{4\}}_+^\times\right\} \cdot$$
$$\cdot Ghm_e e^\pm c \cdot \dim(\ldots) \cdot \Delta_{Exprm.} = \Phi \cdot 10^{99}.$$

And

(306) $$\sqrt[\pi e]{\odot_+^\times \cdot \left\{\pi \frac{\Phi^3\sqrt{i_1 i_2}}{\mathrm{Spin}} \dim_\Phi \Phi^3 i_1 (3+1) \cdot \mathbf{D\{4\}}_+^\times\right\}} \equiv$$
$$\equiv \sin \Delta_{Exprm.}.$$

It is really a finetuned space-time structure, for

(307) $$\odot_+^\times \cdot 137.0359990\ldots = \frac{\mathbf{5202143}}{\mathbf{9}} \cdot 10^{41}.$$

It is unbelievable until after we have a clear picture of cosmological evolution in time

(308) $$\mathbf{5202143}^{G\Pi} = i_1 i_2 \cdot 10^{24}.$$

Finally,

(309) $$\mathbf{5202143}\left\{\frac{Ghm_e e^\pm c \cdot \dim(Ghm_e e^\pm c)}{\alpha a_e}\right\} \equiv 2\Phi^5,$$

to be explained by the reader.

## Fundamental Theorem of Geometry

The following generates both radiation and matter:

(310) $$\frac{\Phi}{i_{1,2}}=\frac{\text{Spin}}{\sin\Delta_{\alpha,\beta}},$$

Consider the first case

(311) $$\frac{\Phi}{i_1}=\frac{\text{Spin}}{\sin\Delta_{\alpha}},$$

where

(312) $$\Delta_\alpha \neq \Delta_{Exprm.}.$$

At first glance, this kind of broken symmetries as if spoils the mathematical beauty of the physical world. But, it is not the case. In contrary, it is very nice to see that actually

(313) $$\frac{\Delta_\alpha}{1+\sqrt[\Phi\pi i_1]{\frac{X}{10^{28}}}}=\Delta_{Exprm.}.$$

In the second case

(314) $$\frac{\Delta_\beta}{\Delta_{Exprm.}}=1+\sqrt[\pi e]{\frac{1}{i_1 i_2\cdot 10^{38}}}$$

The velocity of light can be computed in any manner, including

(315) $$\frac{\otimes^2\Delta_1}{1000}=c+x.$$

Don't bother about the small $x$-correction. No, it does not spoil the picture; in contrary, it is destined to convince you in the fantastical precision of the system of geometry

(316) $$x\cdot e^{5\Phi\pi i_1 e\sqrt{2}}=\mathbf{119298}\cdot 10^{51}.$$

Interpret

(317) $$ci_1=\sqrt{\sqrt{\sqrt{\sin\Theta_W\cdot 10^{69}}}}.$$

Geometry has had at long last its own fundamental theorem.

## Mass and Charge

It is natural that

(318) $$Nme \dim m \cdot e^{\pm} \dim e^{\pm} = \frac{10^{7}}{\Delta_{Exprm.}}.$$

And it should be a very precise configuration for

(319) $$\frac{e^{5\Phi \pi e i_1 \sqrt{2}}}{Nme \dim m \cdot e^{\pm} \dim e^{\pm}} = \mathbf{1048} \cdot 10^{45};$$

(320) $$\mathbf{1048}^{\Phi^3 i_1} = \frac{\mathbf{16945}}{9} \cdot 10^{13}.$$

## Foundations of Alien Technology ?

Technology comes from whatever configurations written once in physics. Beginning with

(321) $$299792458^{\pi} = \mathbf{427285} \cdot 10^{21};$$

(322) $$\sqrt[3+1]{\sqrt[\dim_{\Phi} \Phi^3 i_1]{\mathbf{427285}^{\pi \frac{\Phi^3 i_1}{\text{Spin}}}}} = \frac{10^{9.9999...}}{\Delta_1},$$

we obtain step by step a configuration

(323) $$\left\{ \sqrt[\sin \Delta_{Exprm.}]{\frac{e^{\pm} \dim e^{\pm} \cdot c i_1 \cdot 90 \cdot \Delta_{Exprm.}}{\sin \Delta_{Exprm.}}} \right\}^{\pi\pi} = \mathbf{1427533} \cdot 10^{71}.$$

At least,

(324) $$\mathbf{1427533} \times \mathbf{137} \equiv \pi \frac{\Phi^3 \sqrt{i_1 i_2}}{\text{Spin}}.$$

Compose even a gravi-electroweak configuration (325):

$$\sqrt[\dim_{\Phi} \Phi^3 i_1 \cdot (3+1)]{\left\{ \frac{G}{\Pi} \cdot e^{\pm} \dim e^{\pm} \cdot c i_1 \cdot 90 \cdot \Delta_{Exprm.} \cdot \Theta_W \right\}^{\pi \frac{\Phi^3 i_1}{\text{Spin}}}} =$$

$$= \mathbf{5239} \cdot 10^{7},$$

owing evidently to the fact that

(326) $$\mathbf{5239}^{\pi e} = \mathbf{5772884} \cdot 10^{25} \cdot$$

Denote the previous configuration as *GEW* and now put the world, so to say, on all four so that to see what happens

(327) $$\left\{GEW \cdot 2\Theta_{STR,}\right\}^{\Phi\pi i_1} = \frac{10^{64}}{360};$$

(328) $$GEWS \cdot \pi \frac{\Phi^3\sqrt{i_1 i_2}}{\text{Spin}} \cdot \mathbf{D}\{4\}_+^\times = \frac{10^{56}}{©};$$

(329) $$GEWS \cdot \pi \frac{\Phi^3\sqrt{i_1 i_2}}{\text{Spin}} \cdot \dim_\Phi \Phi^3 i_1 \cdot (3+1) \cdot \\ \cdot \frac{\mathbf{D}\{4\}_+^\times \mathbf{D}\{3\}_+^\times \mathbf{D}\{2\}_+^\times}{\mathbf{InG}_+^\times} = \frac{\mathbf{71864}}{9} \cdot 10^{72}.$$

(330) $$\sqrt[e]{\left\{GEWS \cdot \pi \frac{\Phi^3\sqrt{i_1 i_2}}{\text{Spin}} \cdot \dim_\Phi \Phi^3 i_1 \cdot (3+1)\right\}^{\Phi\pi}} = \\ = \mathbf{212827} \cdot 10^{17}.$$

Note that again

(331) $$\sqrt[\cos 2\Theta_{STR.}]{\mathbf{212827} \cdot 10^{17}} = \mathbf{692316} \cdot 10^{51} \cdot$$

What is remarkable,

(332) $$\left\{GEWS \cdot \dim_\Phi \Phi^3 i_1 \cdot (3+1)\right\} \cdot \\ \cdot \mathbf{InG}_+^\times = \Delta_1 \Theta_W \Theta_{STR.} \cdot 10^{19}.$$

Now it is foreseeable that

(334) $$\frac{GEWS}{a_e} \cdot \frac{\mathbf{10934}}{3} \equiv 1.$$

If write **the unifield configuration** in the form

(335) $$\left\{GEWS \cdot \pi \frac{\Phi^3\sqrt{i_1 i_2}}{\text{Spin}} \cdot \dim_\Phi \Phi^3 i_1 \cdot (3+1)\right\} \cdot \aleph_1 = \mathbf{U},$$

then, geometry will be reduced to the background energy fluctuations as below:

(342)
$$\mathbf{U}\cdot\mathbf{D\{4\}}_{+}^{\times}\cdot\mathbf{InG}_{+}^{\times}=\frac{10^{74.0000...}}{\left\{\left\{\dim E\big/2\right\}^{2}\right\}^{2}}.$$

And necessarily

(343)
$$\left\{\mathbf{U}_{\mathbf{E}}\right\}^{FlucFlucFlucFluc}=\mathbf{4396}\cdot10^{77}.$$

The energy-entropy configuration is justified by

(344)
$$\sqrt[\cos 2\Theta_{STR.}]{\frac{\widehat{E}\breve{E}}{\cos 2\Theta_{STR.}}}\equiv\Delta_{Exprm.}\aleph_{1}\equiv\frac{1}{\alpha a_{e}}.$$

Now we can guess the character of the alien technologies, but we leave this topic for later making here the only remark that the aliens flying their UFOs are desparate to know only one thing: Whether we could pass the cosmological test for intelligence or not. There are some evidences that to them we look ackward but somewhat not quite hopeless. They know that if something is not obtained in due time, those inept civilizations collapse. The cosmic test we may or may not pass is the discovery of the universal invariant, or else, the fundamental metric of geometry, $\Phi$. We have past the test and very probably they know about this event and watch the process on the Globe even more closely. Communications between the cosmic civilizations is possible only in the universal language of Nature which is this geometry. Knowledge and technology transfer is possible only in this language which cannot and should not be taught by the aliens. Civilizations should pass the IQ test on their own and prove their right to exist. If frankly, in the modern theoretical physics one can hardly see any probability of discovery of what we are trying to relate here.

Cosmology is many times more mysterious than we can even imagine. The Solar system energy-entropy resonance state is defined by

(345)
$$\sqrt[Fluc]{\sqrt[Fluc]{\sqrt[Fluc]{\left\{\widehat{E}\breve{E}\cdot\mathbf{U}_{\mathbf{E}}\cdot E\right\}\cdot\odot_{+}^{\times}}}}=\frac{\mathbf{17263}}{9}\cdot10^{13},$$

and to this notably life existence owes to

(340)
$$\mathbf{17263}\cdot\left\{[\![\circledast\ominus]\!]\cdot\mathbf{Cr}_{+}^{\times}\cdot\mathbf{22727}\right\}=\mathbf{290084}\cdot10^{47}.$$

The outer cosmos and the Solar system singularity relation is

(346)
$$\mathbf{290084}\cdot\pi\frac{\Phi^{3}\sqrt{i_{1}i_{2}}}{\text{Spin}}\dim_{\Phi}\Phi^{3}i_{1}(3+1)\cdot$$
$$\cdot\frac{\mathbf{D\{4\}}_{+}^{\times}}{\mathbf{InG}_{+}^{\times}}\mathbf{U}_{\mathbf{E}}=\frac{10^{50}}{\Phi^{3}i_{1}}.$$

## The Organic Life Form

The gene code can be tentatively written as

(347) $$\text{🕸} = \{\mathbf{1,1,5,2,9,2,6,3,9,5,9,5,6,5,6,5,4,1,3,20,61,3}\}_{+}^{\times} = \mathbf{4.434768857 \times 10^{16}}.$$

It tends to comply with some characteristic representations such as

(348) $$\sqrt[\Phi]{\text{🕸}^{\pi e}} = \frac{10^{89}}{\Phi \pi e};$$

(349) $$\text{🕸} \cdot \left\{ \pi \frac{\Phi^3 \sqrt{i_1 i_2}}{\text{Spin}} \cdot \mathbf{D\{4\}_{+}^{\times}} \right\} \cdot \mathbf{U_E} = \frac{10^{75}}{\mathbf{180}};$$

(350) $$\sqrt[e]{\left\{ \text{🕸} \cdot G h m_e e^{\pm} c \cdot \dim(...) \right\}^{\Phi \pi i_1}} = \mathbf{InG_{+}^{\times}} \cdot 10^{37}.$$

What is noticeable,

(351) $$\text{🕸} \cdot \mathbf{23} = 102 \cdot 10^{16}.$$

Taking into account the Denver classification of himan chromosomes, we have a configuration

(352) $$\text{😐} = \{\mathbf{4,7,46,2,3,2,7,3,1,2,2,2,1}\}_{+}^{\times} = \mathbf{212921856.}$$

If most simply, then

353) $$\text{😐} \cdot \Phi^3 i_1 = \mathbf{11473} \cdot 10^5,$$

with an interpretation (354)

$$\mathbf{11473} \cdot \pi \frac{\Phi^3 \sqrt{i_1 i_2}}{\text{Spin}} \cdot \mathbf{D\{4\}_{+}^{\times}} \cdot \mathbf{InG_{+}^{\times}} = \mathbf{17433} \cdot 10^{60}.$$

Evolution is merely SS's growth in complexity up to the human chromosomes. The universal principle is one and the same: the X-shaped intersection of space and time vectors with the Delta angle.

Chromosomes grow into animals subject to hexasymmetry:

(355) $$\left\{ \Phi \Delta_{Exprm.} i_1 \right\}^6 = \frac{10^{21}}{\mathbf{1986567}}.$$

The most advanced breed of primates returns, naturally, back to the fundamental pentagonal symmetry:

(356)
$$\sqrt[\dim_{\Phi}\Phi^3 i_1]{\left\{\left\{\Phi\Delta_{Exprm.}i_1\right\}^5\right\}^{\Phi^3 i_1}} = \frac{10^{21}}{\Delta_1}.$$

So, a unified configuration asks to be written as below:

(357)
$$\left\{\Phi\Delta_{Exprm.}i_1\right\}^6 \times \sqrt[\dim_{\Phi}\Phi^3 i_1]{\left\{\left\{\Phi\Delta_{Exprm.}i_1\right\}^5\right\}^{\Phi^3 i_1}} = \varnothing$$

with the result that

(358)
$$\varnothing^{\Phi} = \frac{10^{60}}{\mathbf{493129}};$$

(359)
$$\sqrt[e]{\varnothing} = \nabla \cdot 10^{10};$$

(360)
$$\sqrt[(3+1)]{\sqrt[\dim_{\Phi}\Phi^3 i_1]{\varnothing^{\pi\frac{\Phi^3 i_1}{\text{Spin}}}}} = \frac{10^{49}}{\Delta_1}.$$

The shortest way to describe biological evolution during the cosmological time span will be

(361)
$$\varnothing = \mathbf{U}_{\mathbf{E}} \cdot i_1 i_2 \cdot 10^{24}\left\{1 + \frac{1}{10000 i_1}\right\}.$$

Biology as it is observed on the planet Earth is necessarily a fact of continuum theory, for

(362)
$$\sqrt[e]{\left\langle \varnothing \cdot \circledast \right\rangle^{\Phi\pi}} = \frac{10^{95}}{\sqrt{\mathbf{163}}},$$

where 163 is the Ramanujan number.

There is, of course, no hope that configurations of human genetics are written right. Nonetheless, they work and look consistent enough. Why so? The general rule is that topological configurations can be written longer or shorter, but not completely wrong, provided only that topological numbers involved are true. Probably, this is the foundation of the reliability of functions and mechanisms of Nature. Accordingly, what we have imagined and wrote in the above is, in principle, not false.

The gene code as modelled by the ultrastring-monochord satisfies conditions of equilibrium in geometry

(363)
$$\sqrt[\Phi]{\frac{\{\mathbf{X}_{\mathbf{M}}, \mathbf{Y}_{\mathbf{F}}\}_{+}^{\times} \cdot \circledast}{\mathbf{\Phi}}} = \tan \mathbf{66} \tan \mathbf{72} \cdot 10^{35}$$

The symmetries of the sixth and fifth order do compete until after the latter wins in evolution. Compare the two schemes:

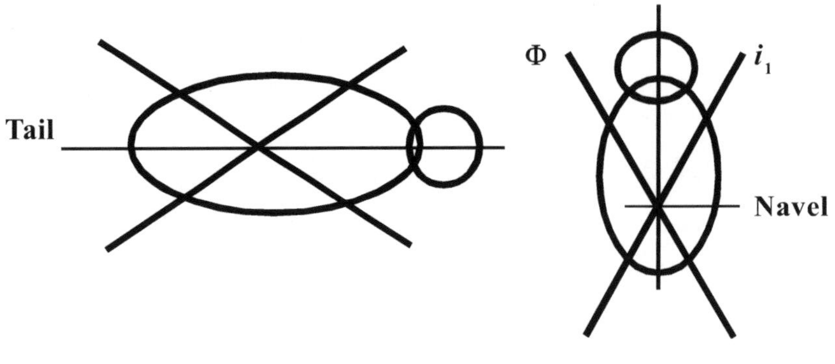


with the Superstructure of absolute geometry:

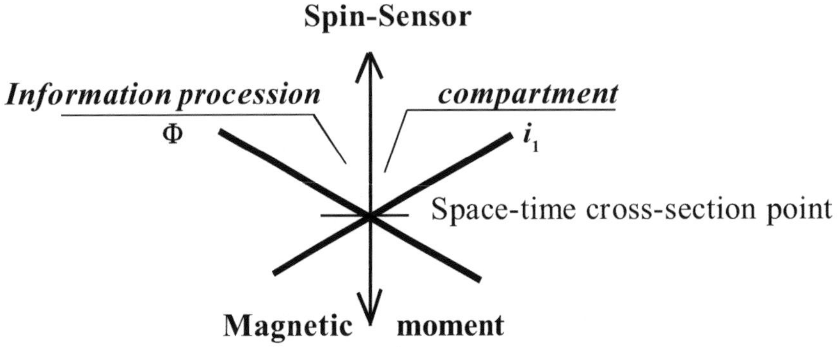


Animals are of hexagonal symmetry and in horisontal projection. Even advanced specii of primates still keep their decorative tails.

But, the SS itself is the upright figure which evolution will tend to. The anthropic evolution begins when apes take vertical position and slowly lose their tails due to the fundamentality of pentasymmetry which in the end takes over.

Yet, this loss should have been compensated according to the gauge principle and it had been compensated very generiously, indeed, and in the following two, that is, bifurcated, ways: notably, by the growth of the brain section in weight and sophistication and, secondly, of the male pride in length and tickness and power.

The final result of evolution is now that man's brain power succeeds to break the universal Golden section code of Existence.

So far so good. Yet, the size of man's male pride envied by the entire animal kingdom has eventually led to quite a controversial result: it is not bad that man is an all four season around sexual animal; yet, the hypersexuality of man implies demographic overproduction. Everything is subject to numeric parameters, and demography is hardly an exclusion. Just tentatively,

$$(364) \qquad \sqrt[e]{\left\langle \mathbf{542981047} \times 6 \cdot 10^9 \right\rangle^{2\Phi\pi i_1}} \equiv \Delta_{Exprm.} \aleph_1,$$

which looks somewhat consistent. If the demographic parameter is exceeded, then Nature will take countermeasures, including AIDS or any other global scale misfortunes which can be easily predicted.

## Harmonicae Mundi

People do never know what they exactly want. Thus, it would be reminded that the end purpose of cognition is the discovery of **the system of world harmony**. If so, here you are. The system is in short something like

(365) $$e^{\{\Phi,\pi,e,i_1\}_+^\times} = \frac{10^{69}}{\mathbf{162}};$$

(366) $$\sqrt[\chi]{e^{\{\Phi,\pi,e,i_1\}_+^\times}} = \mathbf{1547653} \cdot 10^7 \cdot$$

Physics is trivial now, while mathematical biology should be developed from scratch. So, some examples from biology:

(367) $\left\{\mathbf{Cr}_+^\times\right\}^5 \cdot$  $\cdot \{\mathbf{Genes}\}_+^\times = \mathbf{2164031} \cdot 10^{\mathbf{180}};$

(368) $$\mathbf{2164031}^\pi \cdot 3 = \mathbf{2398493} \cdot 10^{14} \cdot$$

Why so, not otherwise? Tie many knots with a long strip of paper. You'll have a helix consisting of regular pentagons. Such is the geometry of the gene code. That's why the fifth power of crystallographic configuration.

Very possibly,

(368) $$Chemy \cdot \sin \Delta_{Exprm.} \sin \Theta_W \cos \Theta_W \cos 2\Theta_{STR.} = \\ = \mathbf{371402} \cdot 10^{13}.$$

Universal mathematical harmony means that everything is explainable and calculable. E.g.,

(369-370) $$\Delta_1 - 137 = U; \\ \pi - 3 = \pi^*;$$

(371) $$\sqrt[U]{\sqrt[U]{e^{\Phi \pi i_1}}} = 77911500000;$$

(372) $$\sqrt[\pi^*]{\pi^* \mathbf{U_E}} = \mathbf{1169715} \cdot 10^{54};$$

(373) $$\sqrt[\pi^*]{\frac{2\Theta_{STR.}}{\cos 2\Theta_{STR.}}} = \exists; \qquad \sqrt[\pi^*]{\Theta_W \Theta_{STR.}} = \frac{2}{3} \cdot 10^{21}.$$

Yet more exercizes

(374-375): $$G^\otimes \equiv \frac{1}{\pi^*}; \qquad h^\oplus \equiv \tan\frac{2\pi}{5};$$

$$\left\langle \frac{\tan 72}{\pi^*} \right\rangle^{4\Phi\pi e i_1} = \mathbf{102466} \cdot 10^{89}.$$

## Superunified Life Equation

Physics is now just dog's game and children's play, as would say the nomads. The everlasting problem is to derive the legitimacy of life from ultimate theoretical principles. Therefore, we have got by the moment the superunified physics and the planetary life crystal so that to obtain

(376)
$$\left\{\frac{Ghm_e e^{\pm}c\cdot\dim(Ghm_e e^{\pm}c)}{\alpha a_e}\right\}\cdot\frac{\cdot\Theta_W\Theta_{STR.}\cdot\{12\times 2\}\cdot\boldsymbol{FB}}{\sin\Delta_{Exprm.}\sin\Theta_W\cos\Theta_W\cos 2\Theta_{STR.}}\cdot\left\{\odot_{+}^{\times}\left\{\left[\left[\frac{\text{🕸}}{\text{😐}}\right]\right]\cdot(\mathbf{Cr}_{+}^{\times}\cdot\mathbf{22727})\right\}\right\}=\frac{\mathbf{4109936}}{9}\cdot 10^{86}.$$

Seems incredible a wishful thinking. But, at least,

(377) $$\mathbf{4109936}^{\chi}=\pi^3\cdot 10^{31.99999...}.$$

Yet, the genuine proof is conceiled in the bottom depth of the universal mathematical machinery

$$\mathbf{4109936}=\frac{\mathbf{274831000}}{2\Theta_{STR.}}.$$

Remind that in the modern sciences life is rated as improbable. The problem is not that of probabilities. The matter is in the teleology of the machinery: to have perfectest possible structures. In this sense cosmology is anthropocentric.

## Fermion-Boson Supersymmetry

(378) $$1+\frac{1}{10}\ln\lg\frac{\otimes^2\cdot\left\langle\Delta_{Exprm.}\Theta_W\Theta_{STR.}\right\rangle\cdot\mathbf{U_E}}{6}=\frac{\Phi\pi}{4};$$

(379) $${}^{Fermion}\frac{\Phi\pi}{4}\leftrightarrow 10\left\{\frac{\Phi\pi}{4}-1\right\}^{Boson};$$

(380) $$\ln\lg\left\{\pi\Phi^3\sqrt{i_1 i_2}\cdot 10^{30}\right\}=Fermion\times Boson.$$

## Delta Parameters

The six Delta operators provide (without comments)

(381) $$\left\{\{\ldots\}_+\right\}^{\chi} = \frac{10^{19.0000000\ldots}}{\mathbf{17077}};$$

(382) $$\left\{\{\ldots\}^{\times}\right\}^{G\Pi} = \frac{10^{51}}{\mathbf{57167}}.$$

(383) $$\{\ldots\}_+^{\times} \cdot \pi \frac{\Phi^3 i_1}{\text{Spin}} \cdot \dim_{\Phi} \Phi^3 i_1 \cdot (3+1) = \frac{\mathbf{1369175}}{9} \cdot 10^{13};$$

(384) $$\mathbf{1369175}^{\Phi\pi e} = \frac{10^{85}}{\Phi}.$$

(385) $$\pi \frac{\Phi^3 \sqrt{i_1 i_2}}{\text{Spin}} \cdot \left\langle \Delta_{\mathbf{Exprm.}} \mathbf{U}_{\mathbf{E}} \right\rangle = \mathbf{605727} \cdot 10^7;$$

(386) $$\mathbf{605727 \times 3.5 \times 4} = \sqrt[\pi]{\mathbf{583749} \cdot 10^{16}}.$$

Or else,

(387) $$\left\langle \mathbf{605727 \times 3.5 \times 4} \right\rangle^{\pi} \equiv \left\{\left\{\dim E/2\right\}^2\right\}^{-2} = \frac{1}{Fluc}.$$

(388) $$\sqrt[\sin \Delta_{\mathbf{Exprm.}}]{E \cdot \mathbf{U}_{\mathbf{E}}} \equiv \frac{10^{21}}{\pi - 3};$$

(389) $$\sqrt[\sin \Delta_{\mathbf{Exprm.}}]{\sqrt[\sin \Delta_{\mathbf{Exprm.}}]{E \cdot \mathbf{U}_{\mathbf{E}}}} \equiv \frac{E_{ag}}{2} \cdot 10^{32}.$$

(390) $$\sqrt[Fluc]{\sqrt[Fluc]{E \cdot \mathbf{U}_{\mathbf{E}} \cdot \widehat{E}\breve{E}}} = e^{\Phi} \cdot 10^{12.00000\ldots}.$$

(391) $$\{\Delta(7)\}_+^{\times} \cdot \circledast \cdot \left\langle E \cdot \mathbf{U}_{\mathbf{E}} \cdot \widehat{E}\breve{E} \right\rangle = \frac{10^{72}}{i_1};$$

(392) $$\sqrt[\sqrt{i_1 i_2}]{\sqrt[Fluc]{\frac{10^{72}}{i_1}}} = \frac{\mathbf{2795}}{3} \cdot 10^{30}.$$

(393) $$\sqrt{i_1 i_2} = T_{Cosm.},$$

(394) $$\sqrt[\sin\Delta_1]{\langle T_{Cosm.}\rangle^{\Delta_1}} = \Phi i_2 \cdot 10^{21}.$$

Sir Newton is said to have warned about the year 2060 of the Christian era, that is,

(395) $$1000\Phi i_2.$$

(396) $$\sqrt[3+1]{\sqrt[\dim_\Phi \Phi^3 i_1]{\langle 1000\Phi i_2 \cdot 365.256\rangle^{\pi \frac{\Phi^3 i_1}{\text{Spin}}}}} = \sqrt[\chi]{\frac{10^{87}}{\mathbf{D\{4\}}_+^\times}}.$$

## Cosmic Time and Pulsations

The only education comes from what goes counter to you.

**Andre Gide**

If pursue the idea of evolution, we may begin with a theorem

(397) $$\left\{\{...\}\frac{\Delta_{Exprm.}}{a_e}\right\}^{T_{Cosm.}\cdot Fluc} \equiv \cos^2 i$$

and blow up it into whatever greater sizes, for example (309),

$$\sqrt[T_{Cosm.}]{\left\{\left\{\{...\}\frac{\Delta_{Exprm.}}{a_e}\right\}\cdot Ghm_e e^{\pm} c\cdot \dim(...)\right\}\cdot \mathbf{D\{4\}}_+^\times \cdot \mathbf{InG}_+^\times} = \exists.$$

To complete the picture, we write

$$\left\{\left\{\{...\}\frac{\Delta_{Exprm.}}{a_e}\right\}\cdot Ghm_e e^{\pm} c\cdot \dim(...)\right\}\Theta_W \Theta_{STR.} = \mathbb{Z}.$$

Then,

(398) $$\mathbb{Z} \equiv e^{\Phi\pi} e^{\Phi\pi i_1}.$$

Now approximate quite a few constants in one go

(399) $$\frac{e^{5\Phi\pi e i_1\sqrt{2}}}{\mathbb{Z}} = \frac{\mathbf{82}}{\mathbf{9}}\cdot 10^{38};$$

(400) $$\left\langle \frac{e^{5\Phi\pi e i_1\sqrt{2}}}{\mathbb{Z}} \right\rangle^{T_{Cosm.}} = \exp e^e \cdot 10^{180-137};$$

(401) $$\sqrt[Fluc]{\left\langle \frac{e^{5\Phi\pi e i_1\sqrt{2}}}{\mathbb{Z}} \right\rangle^{T_{Cosm.}}} = \frac{10^{25.999...}}{a_e}.$$

It is noticeable that

(402) $$e^{5\Phi\pi e i_1\sqrt{2}} \cdot \left\{ T_{Cosm.} \cdot Fluc \right\} = \Phi i_2 \cdot 10^{54}$$

Given the universal $\Phi$-metric of geometry, the Solar system configuration will evolve during the cosmological time span as

(403) $$\odot_{+}^{\times} = \sqrt[T_{Cosm.}]{\frac{10^{57}}{\Phi}}.$$

.

## Cosmological Evolution of Life and Man

> Life is the art of drawing sufficient conclusions from insufficient premises.
> **Samuel butler**

Cosmology is anthropocentric in a sense that vacuum energy fluctuations and the cosmic time rhythm lead to the formation of human genetics:

(404) $$[\![\text{🕸😐}]\!] \cdot \left\{ T_{Cosm.} Fluc \right\} = \Phi\sqrt{i_1 i_2} \cdot 10^{25};$$

(405) $$\left\{ [\![\text{🕸😐}]\!] \cdot \mathbf{Cr}_{+}^{\times} \right\} \cdot \left\{ T_{Cosm.} Fluc \right\} = e^{\Phi\pi} \cdot 10^{42}.$$

The two energy and time related mechanisms do exist in space-time

(406) $$\left\{ T_{Cosm.} Fluc \right\} = \ln\lg\left\{ \pi\Phi^3\sqrt{i_1 i_2} \cdot 10^{30} \right\}.$$

What is crucial,

(407) $$\sqrt[\sin\Delta_{Exprm.}]{\mathbf{U}_{\mathbf{E}} \cdot \frac{\Delta_{Exprm.}}{\sin\Delta_{Exprm.}}} = \sqrt{i_1 i_2} \cdot 10^{17},$$

making universal energy resonance and cosmic time equivalent.

Therefore, no wonder that

(408) $$\left\{ \odot_{+}^{\times} \cdot [\![\text{🕸😐}]\!] \right\} \cdot \left\{ T_{Cosm.} Fluc \right\} = \sqrt{\Delta_\alpha \Delta_\beta} \cdot 10^{68}.$$

The experimental Delta owes to by the self-turbulence of space-time:

(409) $$\frac{\sqrt{\Delta_\alpha \Delta_\beta}}{\Delta_{Exprm.}} = 1 + \sqrt[\Phi^3 i_1]{\pi \frac{\Phi^3\sqrt{i_1 i_2}}{\text{Spin}} 10^{-21}}.$$

## Crystallography: Inert and Animate

Relying upon the fool-proofness of the very method of topological densities, we write the configuration of crystallographic symmetries as

(409) $$\mathbf{Cr}_{+}^{\times} = \{\mathbf{2,13,59,68,36,52}\}_{+}^{\times} \cdot \{\mathbf{230,32,14,6}\}_{+}^{\times} = \\ = 7.830223555 \times 10^{18}.$$

And it succeeds:

(411) $$\left\langle \mathbf{Cr}_{+}^{\times} \right\rangle^{\pi} = \mathbf{22727} \cdot 10^{55};$$

(412) $$\mathbf{22727}^{\pi} \equiv \frac{1}{\Phi i_1}.$$

Now imagine that crystallography has two wings: **inert and animate**. That is,

(413) $$\mathbf{Cr}_{+}^{\times} \times \left\langle \text{🕸☺} \right\rangle.$$

Both wings owe to the evolution in time. Therefore,

(414) $$\sqrt[Fluc]{\left\langle \mathbf{Cr}_{+}^{\times} \times \left\langle \text{🕸☺} \right\rangle \right\rangle^{T_{Cosm.}}} = \exp e^{e} \cdot 10^{26}.$$ Of course,

(415) $$\exp e^{e} \cdot \mathbf{U}_{\mathbf{E}} = \frac{65537}{\Phi^{9}} \cdot 10^{13} \to a_{e};$$

(416) $$\exp\left\langle \frac{1}{10} \lg \frac{\exp e^{e} \cdot E}{10^{10}} \right\rangle = \frac{2}{\Phi}.$$

Double helix crystallization is described by

(417) $$\mathbf{Cr}_{+}^{\times} \cdot e^{2\Phi\pi i_1} = 2\Phi \cdot 10^{24}$$

(418) $$\sqrt[\Phi]{\left\{ \mathbf{Cr}_{+}^{\times} \cdot e^{2\Phi\pi i_1} \cdot \text{🕸} \right\}^{\pi}} = \mathbf{862} \cdot 10^{71.999999...}.$$

Chromosomes are X-shaped following the space-time structure. In case of the human chromosome architecture we have

(410) $$\left\langle \Phi\Delta_{Exprm.} i_1 \times \text{☺} \right\rangle^{\pi} = \frac{10^{36}}{\Delta_1}.$$

The same phenomenon in terms of crystallography:

(419) $$\sqrt[e]{\left\langle \Phi\Delta_{Exprm.} i_1 \times \text{☺} \times \mathbf{Cr}_{+}^{\times} \right\rangle^{\Phi\pi i_1}} = \exp e^{e} \cdot 10^{64}.$$

The synergy of topology and crystallography is such that

$$\sqrt{\sqrt[5]{\pi\frac{\Phi^3\sqrt{i_1 i_2}}{\text{Spin}}\cdot\mathbf{D\{4\}}_{+}^{\times}\cdot\mathbf{Cr}_{+}^{\times}\cdot\circledast}}=\pi\Phi^3 i_1\cdot 10^7 . \quad (420)$$

## Crystallography Plus Morphisms

The system of morphisms studied by Urmantsev as configured

$$\mathbf{Ur}_{+}^{\times}=\mathbf{\{162,192,360,55584,8,255\}}_{+}^{\times}= \\ =7{,}181510295\cdot 10^{19} \cdot \quad (421)$$

Perhaps, it is not accidental that

$$\mathbf{Cr}_{+}^{\times}\cdot\mathbf{Ur}_{+}^{\times}=\sqrt{\sqrt{10^{155}}}\ ; \quad (422)$$

$$\frac{e^{5\Phi\pi i_1 e\sqrt{2}}}{\mathbf{Ur}_{+}^{\times}}\cdot 65537=\frac{10^{36}}{a_e} . \quad (423)$$

Gene code is a product of all possible symmetries:

$$\circledast\cdot\left\{\mathbf{D\{4\}}_{+}^{\times}\cdot\mathbf{Cr}_{+}^{\times}\cdot\mathbf{Ur}_{+}^{\times}\right\}=\frac{10^{100}}{\pi-3}\ ; \quad (424)$$

$$\frac{\widehat{E}\breve{E}}{\pi-3}=\mathbf{180}\cdot 10^{21} . \quad (425)$$

## Computer Simulation of Geometry

Seek simplicity and distrust it.
**Whitehead**

The physical Universe itself is a software which can be restored, copied and stolen. The process will end up with the genuine Artificial Intelligence (AI).

Dirac's monopole shall be in our terms rewritten as

(426) $$\left\langle e^{\pm} e^{\pm}_{ag} \cdot \frac{\Delta_{Exprm.}}{2} \right\rangle = \bigcirc .$$

And what is foreseeable,

(427) $$\cos 2\Theta_{STR.} \sqrt{\cos\Theta_W \sqrt{\sin\Theta_W \sqrt{\sin\Delta_{Exprm.} \sqrt{\bigcirc}}}} \cdot \pi \frac{\Phi^3 \sqrt{i_1 i_2}}{\text{Spin}} = 1 .$$

In addition, an exact

(428) $$\left\langle \frac{10^{23}}{\text{Sing}} \right\rangle^{G\Pi} \cdot \frac{\mathbf{16796}}{3} = 10^{82} .$$

Compute and interpret also

(429) $$\sqrt{\Phi^{\bigcirc}} \equiv \Delta_{Exprm.} \aleph_1 .$$

The above formulae do exist implying that the problem of magnetic monopole can be resolvable. Despite this, the human intelligence will be no more sufficient to develop revolutionary new technologies. In these circumstances humanbeings's purpose will be just to formulate right questions to be put to AI.

UFOs might be a real technology of advanced civilizations, and who knows if they in their times began with (430)

$$\left\langle \pi \frac{\Phi^3 \sqrt{i_1 i_2}}{\text{Spin}} \cdot \dim_\Phi \Phi^3 i_1 \cdot (3+1) \cdot \Delta_{Exprm.} \right\rangle^{\frac{G}{\Pi}} = \frac{10^{63}}{\mathbf{3112879}} .$$

UFOs are superstructures of virtual existence engined by the energy-entropy configuration of the cosmic medium:

(431) $$\frac{\mathbf{3112879}}{\widehat{E}\breve{E}} = \frac{\Delta_1 \Theta_W \Theta_{STR.}}{10^{21}} .$$

Intelligent civilizations seem to have learned first how the Cosmos is grounded on arithmetic:

(432) $$\mathbf{3112879} \times \mathbf{\&}_{+}^{\times} = \Phi\pi \cdot 10^{13}.$$

The next generation of supercomputers for simulation of absolute geometry will not be possible without normal temporature superconducting materials. Anyway,

(433) $$\left\langle e^{\pm} e_{ag}^{\pm} \cdot \Delta_{Exprm.} \right\rangle^{2} \times \mathbf{3112879} \equiv \frac{1}{\mathbf{180}};$$

(434) $$\sqrt[e]{\left\langle \frac{\left\langle E \cdot \widehat{E}\breve{E} \cdot \mathbf{U_E} \right\rangle}{\left\langle e^{\pm} e_{ag}^{\pm} \cdot \Delta_{Exprm.} \right\rangle^{2}} \right\rangle^{\Phi\pi i_1}} = \mathbf{168535} \cdot 10^{71};$$

$$\mathbf{168535} \cdot \mathrm{Sing} \cdot$$

(435) $$\left\langle \mathbf{D\{2\}}_{+}^{\times} \cdot \mathbf{D\{3\}}_{+}^{\times} \cdot \mathbf{D\{4\}}_{+}^{\times} \right\rangle \cdot \mathbf{InG}_{+}^{\times} = \frac{\mathbf{681322}}{3} \cdot 10^{92}.$$

Talks about innovations is self-cheating: resources of existing science and technology are exhausted. The world needs revolutions in technology based upon the Unifield theory, in general, and computer simulations of geometry, in particular. Anyway,

(436) $$\lg \frac{\mathrm{O} \cdot e^{5\Phi\pi i_1 e\sqrt{2}}}{2} = \frac{36000}{e^{\Phi\pi i_1}}$$

## Geometry and Diversions Beyond

Give the method and leave the rest to others' business genius, said the Earl Russell rightly. The method is simple and it has become fairly standard now in our hands. One may begin with the electron and derive immediately a complete exactitude (437):

$$\sqrt[\Phi]{\left\{\frac{\frac{G}{\Pi}\left\{\Phi\Delta_{Exprm.}i_1\right\}\left\{m_e m_{ag} e^{\pm} e^{\pm}_{ag}\right\} ci_1\cdot\Theta_W}{\sin\Delta_{Exprm.}\sin\Theta_W}\right\}^{\pi e}} = \frac{\mathbf{830899}}{3}\cdot 10^{35}.$$

Note that the experimental values made of use in this work appear to be of the best possible accuracies. Secondly, the integer, 830899, looks suspectibly too ordinary. But, it is a very clever number, e.g.,

(438) $$\mathbf{830899}\cdot\mathrm{Sing}\cdot\dim_\Phi \Phi^3 i_1\cdot(3+1) = \mathbf{227489000}.$$

Indeed, what matters is which way you write space-time (439):

$$\sqrt[\dim_\Phi \Phi^3 i_1\cdot(3+1)]{\left\{\frac{G}{\Pi}\left\langle\Phi\Delta_{Exprm.}i_1\right\rangle\cdot \left\langle m_e m_{ag} e^{\pm} e^{\pm}_{ag}\right\rangle\cdot ci_1\right\}^{\pi\frac{\Phi^3 i_1}{\mathrm{Spin}}}} = \sqrt[\chi]{\mathbf{287087}\cdot 10^{35}}.$$

And,

(440) $$\mathbf{287087}^{\Phi\pi i_1} \equiv \pi\frac{\Phi^3\sqrt{i_1 i_2}}{\mathrm{Spin}}.$$

Compute and interpret (441-444)

$$\frac{\left\langle \pi\frac{\Phi^3\sqrt{i_1 i_2}}{\mathrm{Spin}}\cdot\frac{G}{\Pi}\cdot m_e m_{ag} e^{\pm} e^{\pm}_{ag} = \Xi\right\rangle^{\pi}}{1-\frac{\Phi\pi e\cdot\mathrm{Sing}}{10^6}} = 2\Phi\cdot 10^{12};$$

$$\Xi^{\pi\pi} \equiv 2; \qquad \left\langle \Xi\cdot\Delta_{Exprm.}\cdot ci_1\right\rangle^{\pi} = \mathbf{112}\cdot 10^{19};$$

$$\left\langle \Xi\cdot\dim_\Phi \Phi^3 i_1\cdot(3+1) = \Sigma\right\rangle^{2\chi} = \mathbf{8636994}\cdot 10^{45};$$

(445) $$\frac{\left\langle \Xi \cdot \dim_{\Phi} \Phi^3 i_1 \cdot (3+1) \cdot \Delta_{Exprm.} \cdot ci_1 = \mathrm{M} \right\rangle^{\pi}}{\sin \Delta_{Exprm.}} = \exp \Phi^5 \cdot 10^{20};$$

(446) $$\frac{\left\langle \mathrm{M} \cdot \Theta_W \right\rangle^{\pi}}{\sin \Delta_{Exprm.}} \equiv \pi \frac{\Phi^3 i_1}{\mathrm{Spin}};$$

(447) $$\Phi \Delta_{Exprm.} i_1 \cdot m_e m_{ag} e^{\pm} e^{\pm}_{ag} \cdot ci_1 = \sqrt[\pi]{\frac{10^{60}}{\mathbf{D\{4\}}^{\times}_{+}}}.$$

The electron well equipped with qualities shall be

(448) $$\frac{\frac{G}{\Pi} \left\langle \Phi \Delta_{Exprm.} i_1 \right\rangle \left\langle m_e m_{ag} e^{\pm} e^{\pm}_{ag} \right\rangle ci_1 \cdot \Theta_W}{\sin \Delta_{Exprm.} \sin \Theta_W} = \mathrm{K}.$$

Now project it into the 4-Dim topology and see (355)

(449) $$\mathrm{K} \cdot \pi \frac{\Phi^3 \sqrt{i_1 i_2}}{\mathrm{Spin}} \left\langle \mathbf{D\{2\}}^{\times}_{+} \cdot \mathbf{D\{3\}}^{\times}_{+} \cdot \mathbf{D\{4\}}^{\times}_{+} \right\rangle \equiv \mathbf{U_E}.$$

If so, then be prompt to derive

(450) $$\sqrt[e]{\left\langle \frac{\exists}{©@} \cdot \mathbf{U_E} \cdot \mathbf{InG}^{\times}_{+} \right\rangle^{\Phi\pi}} = \exists.$$

But, this in turn immediately urges to finalize the electron:

(451) $$\frac{\sqrt[e]{\left\langle \frac{\exists}{©@} \cdot \mathbf{U_E} \cdot \mathbf{InG}^{\times}_{+} \right\rangle^{\Phi\pi}} \equiv \exists}{\mathbf{D\{4\}}^{\times}_{+}} \equiv \frac{1}{\Phi}.$$

Anyway, electromagnetism written as (452)

$$\frac{G}{\Pi} \left\langle \Phi \Delta_{Exprm.} i_1 \right\rangle \left\langle e^{\pm} e^{\pm}_{ag} \right\rangle ci_1 \times e^{2\pi\Phi i_1} = \pi \Phi^3 i_1 \cdot 10^{9.000000...}$$

generates Faraday's lines of forces.

Both classical and quantum electrodynamics reduce to such expressions as (452)

$$\sqrt[\mathrm{Spin}]{\frac{G}{\Pi} \left\langle \Phi \Delta_{Exprm.} i_1 \right\rangle \left\langle e^{\pm} e^{\pm}_{ag} \right\rangle ci_1 \times e^{2\pi\Phi i_1}} = 4\Phi \cdot 10^{11};$$

(453) $$\left\langle \frac{G}{\Pi} \cdot \frac{\left\langle \Phi\Delta_{Exprm.} i_1 \right\rangle \left\langle e^{\pm} e^{\pm}_{ag} \right\rangle c i_1}{\sin \Delta_{Exprm.}} \cdot \frac{1}{a_e} \cdot \mathbf{U_E} \right\rangle^{\pi} = \left\langle \mathbf{13829} \cdot 10^{29} \right\rangle^{\Phi}.$$

The process proves to be golden-algorithmic up to the end

(454) $$\mathbf{13829}^{\Phi\pi} = \frac{10^{21.0000...}}{\cos \Theta_W}.$$

Gravi-electromagnetism in a concise form will be

(455) $$\left\langle \lg \left\langle \left\langle \frac{G}{\Pi} \cdot \pi \frac{\Phi^3 \sqrt{i_1 i_2}}{\text{Spin}} \cdot \mathbf{D\{4\}}^{\times}_{+} \right\rangle \cdot 2 N m_e m_{ag} \cdot e^{\pm} e^{\pm}_{ag} \right\rangle \right\rangle^{\chi} =$$
$$= \Phi^3 i_1 \cdot 10^8.$$

Newton's law of universal gravitation has to be put as

(456) $$\frac{G}{\Pi} \left\{ \frac{N m_e m_{ag}}{\Phi^2} \right\}^2.$$

This is one of key moments when geometry shifts to physics and, therefore, we should look for very basic but highly nontrivial solutions such as an incredibly accurate

(457) $$\exists \left\{ \frac{G}{\Pi} \left\{ \frac{N m_e m_{ag}}{\Phi^2} \right\}^2 \right\}^{\exists} = \frac{9}{\mathbf{6641569}} \cdot 10^{64}.$$

In the given case

(458) $$4 \left\{ 1 + \frac{1}{10} \ln \lg \frac{\sqrt[e]{\mathbf{6641569}^{\Phi \pi i_1}}}{\pi \Phi^3 i_1} \right\} = \Phi\pi.$$

In the unifield theory takes place the following bifurcation:

(459) $$\underset{\textit{Fermion Section}}{\left\{ \frac{\Phi\pi}{4} = \frac{\text{Spin}}{\sin\Delta_\beta} \right\}} \Leftrightarrow 10 \underset{\textit{Boson Section}}{\left\{ \frac{\Phi\pi}{4} - 1 \right\}}.$$

The experimental Delta value arises owing to the self-perturbation of the entire geometry

(460) $$\Delta_{Exprm.} = \Delta_\beta - \pi \frac{\Phi^3 \sqrt{i_1 i_2}}{\text{Spin}} \frac{\mathbf{D\{4\}}^{\times}_{+}}{\mathbf{12054} \cdot 10^{45}};$$

(461) $$\lg\lg\frac{\mathbf{12054}^{\Phi\pi e i_1}}{\Phi^3 i_1}=\frac{1}{\Pi_{Cosm.}}=\dim G_{Newton}.$$

Analyse (462-463)

$$\sqrt[U]{\frac{G}{\Pi}\left\{\frac{Nm_e m_{ag}}{\Phi^2}\right\}^2\cdot\left(1-\sqrt[G\Pi]{\frac{1}{526122\cdot 10^{13}}}\right)}=2\pi\cdot 10^{16};$$

$$\frac{G}{\Pi}\left\{\frac{Nm_e m_{ag}}{\Phi^2}\right\}^2\cdot\exists^2=\frac{10^8}{\aleph_1},$$

or even

(464) $$\frac{G}{\Pi}\left\{\frac{Nm_e m_{ag}}{\Phi^2}\right\}^2\cdot\frac{1618449\cdot 10^{40}}{\mathbf{D\{4\}}_+^\times}=\frac{10^8}{\aleph_1}.$$

And naturally,

(465) $$\sqrt[e]{\left\langle\frac{G}{\Pi}\left\{\frac{Nm_e m_{ag}}{\Phi^2}\right\}^2\cdot\mathbf{U_E}\right\rangle^{2\Phi\pi i_1}}=\ln\frac{2}{\Phi}\cdot 10^{90};$$

(466) $$\frac{\exists}{©@}\cdot\frac{G}{\Pi}\left\{\frac{Nm_e m_{ag}}{\Phi^2}\right\}^2\cdot E=\frac{\mathbf{689}}{3}\cdot 10^{14}.$$

The energy-entropy bigbang scenario will be

(467) $$\frac{\sqrt[e]{\left\{\frac{G}{\Pi}\left\{\frac{Nm_e m_{ag}}{\Phi^2}\right\}^2\cdot\widehat{E}\breve{E}\right\}^{\Phi\pi i_1}}}{1+\frac{\Phi}{1000}}=(©@)^2\cdot 10^{77}.$$

What is noticeable (468),

$$\frac{G}{\Pi}\left\{\frac{Nm_e m_{ag}}{\Phi^2}\right\}^2\cdot\sin\Theta_W=10000\left\langle\Delta_{Exprm.}\aleph_1\right\rangle\left\langle 1+\frac{\Phi^3 i_2}{10000}\right\rangle.$$

Newton's and Einstein's concepts of gravity can and must be unified, e.g., (469-471):

$$\frac{G}{\Pi}\left\{Nm_e m_{ag}\right\}^2 \cdot \left\{\pi \frac{\Phi^3 \sqrt{i_1 i_2}}{\text{Spin}} \mathbf{D\{4\}}_+^\times\right\} \cdot \mathbf{U_E} = \mathbf{867} \cdot 10^{63};$$

$$\mathbf{867}^{\pi e} = \mathbf{123} \cdot 10^{23};$$

$$\frac{G}{\Pi}\left\{Nm_e m_{ag}\right\}^2 \cdot \left\{\pi \frac{\Phi^3 \sqrt{i_1 i_2}}{\text{Spin}} \frac{\mathbf{D\{4\}}_+^\times}{\mathbf{InG}_+^\times}\right\} \cdot \mathbf{U_E} = \sqrt{10^{105}}.$$

All this is probable and there is nothing to wonder about, but one of improbabilities of cosmology is the gravitational configuration of the Solar system as follows (472):

$$\odot_+^\times \cdot \left\{\frac{G}{\Pi}\left\{\frac{Nm_e m_{ag}}{\Phi^2}\right\}^2 \cdot \left\{\pi \frac{\Phi^3 \sqrt{i_1 i_2}}{\text{Spin}} \cdot \mathbf{D\{4\}}_+^\times\right\}\right\} = \frac{10^{101}}{\Phi}.$$

Thus, we have to finally obtain (473-474)

$$\frac{\dim_\Phi \Phi^3 i_1 \cdot (3+1) \times \mathbf{U_E}}{\Phi} = \pi \frac{\Phi^3 \sqrt{i_1 i_2}}{\text{Spin}} \cdot 10^9;$$

$$\odot_+^\times \cdot \left\{\frac{G}{\Pi}\left\{\frac{Nm_e m_{ag}}{\Phi^2}\right\}^2 \cdot \left\{\begin{matrix}\dim_\Phi \Phi^3 i_1 \cdot (3+1) \cdot \\ \cdot \mathbf{D\{4\}}_+^\times\end{matrix}\right\}\right\} \mathbf{U_E} = 10^{110}.$$

As Sir Isaac Newton did foresee, the Solar system proves to be, indeed, a most perfectly fine tuned clock work.

(The only misery is that the previous calculations contradict to the IAU's recent resolution concerning the fate of Pluto. As far as I should be pretending for certain major international money prizes in sciences, I cannot but feel uneasy about in these circumstances and therefore I would rush to agree with that Pluto is not that small but brave outernmost guardian of the stability of the Solar system. At the same time, I would argue that Pluto still orbits the Sun and as a sovereign celestial body it is free to be accounted for in calculations.)

Whatsoever, the Solar system gravitational space-time configuration is an extraordinary phenomenon. It is an inverse function of the $\Phi$-metric designed, notably, to focus the cosmic universal energy flow on the Earth planet:

(475-478)

$$\frac{G}{\Pi}\left\{Nm_e m_{ag}\right\}^2 \cdot \mathbf{U_E} = \sqrt[\pi]{2 \cdot 10^{60}};$$

$$\frac{G}{\Pi}\left\{Nm_e m_{ag}\right\}^2 \cdot \mathbf{U_E} \cdot \odot_+^\times = \frac{10^{65}}{e^e};$$

$$\frac{\mathbf{D\{4\}}_{+}^{\times}}{e^{e}} = \mathbf{187} \cdot 10^{42};$$

$$\pi \frac{\Phi^{3}\sqrt{i_1 i_2}}{\text{Spin}} \left\{ \dim_{\Phi} \Phi^{3} i_1 \cdot (3+1) \right\} \frac{\mathbf{D\{4\}}_{+}^{\times}}{e^{e}} = \frac{1}{\mathbf{19532}} \cdot 10^{51}.$$

Consequently, we do eventually arrive at what should be called geo- and anthropocentric cosmology. The Earth-centric Solar system configuration and the gene code self-organization of matter are indispensible through such great but hidden workings of the universal mathematical machinery as (479-481)

$$\frac{\text{🕸}}{\Phi} = 1000\mathbf{InG}_{+}^{\times};$$

$$\frac{\text{☺}}{\Phi} = \frac{\aleph_1 \cdot 10^{10}}{65537};$$

$$\frac{[\![\text{🕸☺}]\!]}{\Phi} = \frac{10^{24.99999...}}{\left\{ \left\{ \frac{\dim E}{2} \right\}^{2} \right\}^{2}},$$

where the problem reduces to the background energy fluctuations which notably does permanently stir the rotten since long stagnant waters of the Cantor's continuum considered, in addition, through the narcotized eyes of diehard axiomatic formalism.

Also (482):

$$\frac{[\![\text{🕸☺}]\!]}{\Phi} \left\{ 1 + \frac{\mathbf{Sing} \cdot \mathbf{D\{4\}}_{+}^{\times} \cdot \mathbf{InG}_{+}^{\times} \cdot \mathbf{U}_{\mathbf{E}}}{\mathbf{255472} \cdot 10^{69}} \right\} = \frac{10^{25}}{\left\{ \left\{ \frac{\dim E}{2} \right\}^{2} \right\}^{2}};$$

(483) $$\mathbf{255472}^{\Phi^{3} i_1} = \mathbf{137} \cdot 10^{27.00000...};$$

(484) $$\mathbf{255472}^{\pi \Phi^{3} i_1} = \frac{10^{93.9999...}}{180\Phi}.$$

The cause for the background energy fluctuations is the constant bifurcation-oscillation of space and time vectors:

(485) $$\mathbf{137}^{\pi \frac{\Phi^{3} i_1}{\text{Spin}}} \cdot \left\{ \left\{ \frac{\dim E}{2} \right\}^{2} \right\}^{2} \equiv 1;$$

(486) $$\sqrt[Fluc]{\sqrt[Fluc]{E\cdot\mathbf{U}_{\mathbf{E}}\cdot\widehat{E}\breve{E}}}=\frac{10^{38}}{e^{\Phi}}=e^{\Phi}\cdot10^{12}.$$

Then, the cosmological equilibrium shall certainly be

(487) $$\sqrt[(3+1)]{\dim_{\Phi}\Phi^{3}i_{1}\sqrt[\pi\frac{\Phi^{3}i_{1}}{\mathrm{Spin}}]{\mathbf{137}}}=\sqrt[G\Pi]{©\cdot10^{11}}\cdot$$

It is crucial to conceive the idea that the Superstructure of geometry, no matter how senseless it may look to some too much wiser ones, is nevertheless the Holy Womb of Nature. It is that mythological Magical Mirror which every other nation on Earth dreamed to have to learn the secrets of the Universe. If you don't mind, it is the Philosophical Stone legended since the dawn of history both in the Oxident and the Orient. Some samples of workings inside the Superstructure of space-time and matter can be displayed:

(488) $$\left\{\frac{\Delta_{Exprm.}}{\sin\Delta_{Exprm.}}\right\}^{\pi\Phi^{3}i}=\frac{\mathbf{293912}}{3}\cdot10^{34};$$

(489) $$\sqrt[\sin\Delta_{Exprm.}]{\sin\Delta_{Exprm.}\sqrt{\sin\Delta_{Exprm.}\sqrt{\Theta_{W}\Theta_{STR}}}}=\Phi\frac{4}{\pi}\cdot10^{9}.$$

It is no less crucial to understand that in the bottom depth of reality do necessarily rule trigonometric constants alone. This is simply by the reason that only trigonometric parameters can be absolute. The Existence would need some absolute foundations. In this respect it should be pointed out that the ideology of relativity makes everything rotten, including morality. Anyway, in the unifield theory we have found not many but four trigonometric functions (490):

$$\sin\Delta_{Exprm.}\sin\Theta_{Weinberg}\cos\Theta_{Weinberg}\cos2\Theta_{Strong\ nuclear\ force}=\mathrm{Trig}.$$

This configuration works as follows below (491-499):

$$\sqrt[\mathrm{Trig}]{\frac{\Delta_{Exprm.}\Theta_{W}\Theta_{STR}}{\mathrm{Trig}}}=\frac{10^{102}}{\mathbf{D\{4\}}_{+}^{\times}};$$

$$\sqrt[\mathrm{Trig}]{\Theta_{W}\Theta_{STR}}=\frac{10^{34}}{\mathbf{2928843}};$$

$$\sqrt[\mathrm{Trig}]{\frac{2\Theta_{STR}}{\cos2\Theta_{STR}}}=\sin\Delta_{Exprm.}\cdot10^{21};$$

$$\sqrt[\mathrm{Trig}]{\aleph_1} = \frac{10^{30}}{\mathbf{386}}; \qquad \sqrt[\mathrm{Trig}]{\mathbf{U}_{\mathbf{E}}} = \frac{10^{87}}{\sin\Theta_W};$$

$$\frac{E\cdot\mathbf{U}_{\mathbf{E}}\cdot\overset{\frown}{E}\breve{E}}{\mathrm{Trig}} = \frac{10^{38}}{\Pi_{Cosm.}};$$

$$\frac{\odot^{\times}_{+}}{\text{🕸}} = \mathrm{T}\times\Theta_W\Theta_{STR}\cdot 10^{26}; \qquad \sqrt[\mathrm{Trig}]{\mathbf{542981047}} = \cos^3 i\cdot 10^{81}.$$

**The Solar system planetary life logos** has to be determined by the strictest possible reasons, indeed,

(500) $$\frac{\left\{\odot^{\times}_{+}\cdot[\![\text{🕸}\text{😐}]\!]\right\}}{\mathrm{Trig}} = \mathbf{371922}\cdot 10^{65}$$

The twin-testicle operators of Nature will certainly pretend for the same result of conception of life:

(501) $$\frac{\odot^{\times}_{+}}{\text{🕸}\text{😐}}\cdot\mathrm{Trig}^2 \equiv \{©@\}^2.$$

The proof lies as ever in the character of symmetry violations

(502): $$\frac{\odot^{\times}_{+}}{\text{🕸}\text{😐}} = \frac{\{©@\}^2}{\mathrm{Trig}^2}\cdot 10^{19}\left\{1+\left\{\left\{\frac{3}{11\cdot 10^{22}}\right\}^{\cos 2\Theta_{STR}}\right\}^{\cos 2\Theta_{STR}}\right\},$$

assuring the consistence of the whole picture. Mathematics' purpose is to name things by their proper names. That's why we name testicle operators as testicle operators. In addition, we are as coward as anyone else on Earth and we do always reserve some finally knocking down arguments:

(503) $$\left[\!\left[\left[\!\left[\odot^{\times}_{+}\cdot\text{🕸}\text{😐}\right]\!\right]^{\mathrm{Trig}}\right]\!\right]^{\mathrm{Trig}} = 2\pi.$$

The trigonometric constants are so powerful simply because they run energy in all its forms.

Geometry is one for all universes. The quintuplet of universal physical constants composes a configuration. So, immediately,

(504) $$\sqrt[\mathrm{Trig}]{Ghm_e e^{\pm}c\cdot\dim(Ghm_e e^{\pm}c)} = 2\pi Rad\cdot 10^{40}.$$

All the known physics arrives at

(505) $$\sqrt[\mathrm{Trig}]{\frac{\mathrm{Quintuplet}}{\alpha_{Sommerfeld}a_{Elecron}}\cdot\mathrm{Trig}} = \frac{10^{83}}{\mathbf{137}}.$$

Now we have to very neatly soft land on with no harm to anydody's honourable reputation for which purpose we derive all the known and unknown physics at one go as follows below (506):

$$\frac{\left\| \frac{Ghm_e e^{\pm} c \cdot \dim(Ghm_e e^{\pm} c)}{\alpha_{Sommerfeld} a_{Elecron}} \cdot \pi \frac{\Phi^3 \sqrt{i_1 i_2}}{\text{Spin}} \dim_{\Phi} \Phi^3 i_1 \cdot (3+1) \cdot \mathbf{D}\{4\}_{+}^{\times} \right\|}{\text{Trig}} = \cos\frac{2\pi}{5} \cdot 10^{59.00000...}.$$

Yet again an exact

(507) $$\star \cdot \text{Trig} = \frac{\mathbf{14247436}}{3} \cdot 10^{9}.$$

To the attentive and benevolent reader it is already clear that the Unifield theory is nothing but termodynamics of continuum now discretized. Thus, the Boltzman constant has to be a purist geometric phenomenon

(508) $$\sqrt[k_{Bolzmann}]{\left\{ \pi \frac{\Phi^3 \sqrt{i_1 i_2}}{\text{Spin}} \dim_{\Phi} \Phi^3 i_1 \cdot (3+1) \cdot \Delta_{Exprm.} \right\} \cdot \mathbf{D}\{4\}_{+}^{\times}} = \left\{ \left\{ \frac{\dim E}{2} \right\}^2 \right\}^2 \cdot 10^{36},$$

whence the theoretical value of termodynamical constant is

(509) $$k_{Boltzman} = 1.380658328.$$

Consider such expressions and likes (510-513)

$$\frac{\mathbf{U_E}}{k^2} = \frac{10000}{\alpha a_e}; \qquad \left\{ E \cdot \mathbf{U_E} \cdot \widehat{E}\breve{E} \right\}^k = \frac{10^{53}}{32};$$

$$E^{kkkk} \equiv \frac{\aleph_1}{65537}; \qquad k^{137.000...} = \frac{10^{22}}{e^{\Phi \pi i_1}}.$$

Geometry's purpose is to fill the cosmic void with anything worthy, including the nonlocal point-singularity (with the previous $k$)(514)

$$\sqrt[\mathbf{0.7797}]{\sqrt[\mathbf{0.7408}]{\left\{ \left\{ \pi \frac{\Phi^3 \sqrt{i_1 i_2}}{\text{Spin}} \dim_{\Phi} \Phi^3 i_1 \cdot (3+1) \right\}^{\pi e} \right\}^{k}}} = \mathbf{568} \cdot 10^{47},$$

or human chromosomes

(515) $$\sqrt[Rogers']{\sqrt[Kepler's]{😐}} = \frac{10^{21}}{\exp e^{e}},$$

or even humanbeings themselves (517-519)

$$\sqrt[Kepler's]{\mathbf{542891047}} = \frac{10^{12}}{\Phi};$$

$$\sqrt[Kepler's]{\left\{\mathbf{542891047} \times 10^{9}\right\}^{Rogers'}} = \sqrt[kk]{\exp e^{e} \cdot 10^{29}};$$

$$\sqrt[Kepler's]{\left\{\mathbf{542891047} \times 6 \cdot 10^{9}\right\}^{Rogers'}} = \sqrt[kk]{\Delta_{0} \cdot 10^{35}}.$$

Consider the double helix, that is, torsional, structure defined by the *apriori* equations such as

(520) $$\frac{🕸}{\text{Spin}} = \left\{©@\right\}^{2} \cdot 10^{18};$$

(521) $$\sqrt[\text{Spin}]{\sqrt[\text{Spin}]{\sqrt[\text{Spin}]{\sqrt[\text{Spin}]{\sqrt[\text{Spin}]{\frac{🕸}{\text{Spin}}}}}}} = 2 \cdot 10^{34}.$$

By the very algorithmicity of Euclid-Gaussian geometry the gene code topology cannot be otherwise than (522)

$$\sqrt[\text{Spin}]{\sqrt[\text{Spin}]{\{\mathbf{3,5,17,257,65537}\}_{+}^{\times} \cdot 🕸}} = 180\Phi \cdot 10^{39}.$$

It should be remarked that

(523) $$\sqrt[0.7408 \times 0.7797]{\frac{\Delta_{Exprm.}}{\alpha}} = \sqrt[kk]{\frac{\Phi\pi}{4} \cdot 10^{14}},$$

where there is no tautology of Delta.

## Goedel's Theorem and Consequences

Arm in Arme, hoeher stets und hoeher,
Vom Mogolen bis zum griech'schen
Seher...
**Schiller**

The two mythological code numbers come from nomadic folklore and written sources attributed to Hipparchus, correspondingly. They are, indeed, prophetic:

$$\sqrt[\text{Trig}]{\mathbf{1292049}} \equiv \cosh\cosh\cos^2 i;$$

(524-525) $$\sqrt[\text{Trig}]{\mathbf{310952}} \equiv \pi\frac{\Phi^3 i_1}{\text{Spin}}.$$

And, of course (526-527),

$$\left\{\cosh\cos^2 i\cdot\pi\frac{\Phi^3 i_1}{\text{Spin}}\right\}e^{\Phi\pi i_1} = \frac{\mathbf{616882}}{\mathbf{9}};$$

$$\frac{\mathbf{616882}}{\mathbf{9}}\{\text{🕸😐}\} \equiv 4\Phi \equiv \frac{\mathbf{InG}_+^\times}{\Phi^3},$$

*et cetera ad infinitum*. And, of course,

(528) $$\sqrt[e]{\{\mathbf{Nom,Hip}\}_+^{\times\Phi\pi}} = \mathbf{2008}\cdot 10^{30}$$

The key point is that the Unifield theory rounds up with scientific rationality and therefore, as a consequence of Goedel's theorem, we come in touch with the irrational section of the cosmological reality, the latter being independent from our will. Scientists tend to faint at any reason as medieval European court ladies, but they should not to do so before the infinity of cosmological reality. Fundamentalism even in scientific formalism is not good, according to Goedel's famous theorem.

It is one thing that the code numbers provide a rational

(529) $$\{\mathbf{Hi,N}\}_+^\times \times \left\langle \pi\frac{\Phi^3\sqrt{i_1 i_2}}{\text{Spin}}\cdot\dim_\Phi \Phi^3 i_1\cdot(3+1)\right\rangle = \frac{10^{20}}{\copyright}.$$

But, there is yet another point. We have calculated man's name in the unifield theory coming to the number, **542891047**. The Bible says that one of man's names is, 666, and in this capacity Homo Sapiens will once reveal himself as the most fearful beast. Anyway, by the moment we have become intelligent enough to check whether man's name is eventually 666, or not. The universal Greek-Mongol code is an revelation, too.

Therefore,

(530) $$\left\{\frac{\mathbf{\{Hi, N\}_{+}^{\times}}}{\mathbf{542891047}\times 666}\right\}^{\chi}\cdot 2\pi Rad = \Phi\cdot 10^{34}$$

(531) $$34^{\chi\chi}\cdot\frac{\mathbf{482198}}{9} = 10^{44};$$

(532) $$\mathbf{482198}^{2\Phi\pi i_1} = \mathbf{312415}\cdot 10^{68}\cdot$$

Science and religion are the two wings of the cognitive process. Both are based on faith. But they differ as knowledge and belief differ. Despite this, the unity of cosmological reality presupposes the unity of science and religion in some final scheme.

Religions are possible if only the concept of the soul can be validated by scientific methods. The key problem of cognition is eternally the hypothesis of soul. There are two consequetive dualities: first, of body and mind, and secondly, of brain and phyche.

The brain is the most complex hardware system composed of all the topological configurations we now know or we will know in the future. The system records all the individual human experience. What happens to this information content after death? Energy and information look somewhat equivalent implying that every lifelong accumulated experience can be conserved. But how? In the unifield theory we describe the only object, notably, the mathematical point, for what does only exist is the mathematical point-singularity. The brain as a hardware is a singular structure of which formula can be, in principle, written. At the same time this formula will be an abstraction embracing however the point-singularity as a whole with its practically infinite degrees of freedom, that is, memory cells and connexions between them can record and conserve any amount of information. Consequently, formulae for the human brain and for the soul are one and the same implying that the information content of the brain equal to an individual experience may continue its existence in the cosmic void.

**Nur die Fuelle fuehrt zur Klarheit**
**Und im Abgrund wohnt die Wahrheit,**

says Schiller meaning that only completeness leads to clarity, and truth resides in the very depth. Anyway, we shall look for the formula for the soul, if any, of course.

To begin with, remind that there are two anomalies that define boundaries of both our knowledge and ignorance:

(533) $$\left\{\Delta_{Exprm.}\aleph_1 = \frac{1}{\alpha a_e}\right\}^{\oplus} = \mathbf{1783}\cdot 10^{43}.$$

This is legitimate because of (534)

$$\mathbf{1783}\cdot\pi\frac{\Phi^3\sqrt{i_1 i_2}}{\text{Spin}}\cdot$$

$$\frac{\cdot\left\langle \mathbf{D\{2\}}_+^\times\cdot\mathbf{D\{3\}}_+^\times\cdot\mathbf{D\{4\}}_+^\times\right\rangle\cdot\mathbf{U_E}\cdot\mathbf{InG}_+^\times}{10^{105}}+1=\cos i.$$

The self-gravitating space-time in complete is (535)

$$\sqrt[\Phi]{\left\langle \begin{matrix}\frac{\exists}{©@}\cdot\pi\frac{\Phi^3\sqrt{i_1 i_2}}{\text{Spin}}\dim_\Phi\Phi^3 i_1\cdot(3+1)\cdot\left\langle\Delta_{Exprm.}\aleph_1\right\rangle\cdot\\ \cdot G\dim G\end{matrix}\right\rangle^{\pi e}}=$$

$$=2\pi\cdot10^{52}.$$

We may reduce the physical Universe to (536)

$$\frac{\mathbf{10^{114}}}{\mathbf{55}}=\frac{\exists}{©@}\pi\frac{\Phi^3\sqrt{i_1 i_2}}{\text{Spin}}\dim_\Phi\Phi^3 i_1\cdot(3+1)\left\langle\Delta_{Exprm.}\aleph_1\right\rangle\cdot$$

$$\cdot\left\langle \mathbf{D\{2\}}_+^\times\cdot\mathbf{D\{3\}}_+^\times\cdot\mathbf{D\{4\}}_+^\times\cdot\mathbf{InG}_+^\times\right\rangle\cdot\mathbf{U_E}\cdot Ghm_e e^\pm c\cdot\dim(...).$$

It is believable thanks to the existence of quite an exact

(537) $$\mathbf{55}^{\chi\cdot\frac{G}{\Pi}}=\mathbf{58527949}\cdot10^{24}\cdot$$

Assuming that the inert-animate crystallography is subject to a general configuration

(538) $$\left\langle\mathbf{Cr}_+^\times\cdot\circledast\right\rangle,$$

it is easy to find out that

(539) $$\mathbf{Cr}_+^\times\cdot2\Theta_{STR.}=E_{ag}^2\cdot10^{20};$$

(540) $$\left\langle\mathbf{Cr}_+^\times\cdot\circledast\right\rangle\cdot\mathbf{55}\equiv E_{ag}^{-2}.$$

Now that the organic life form in the Solar system of nine planets asks to be written as (541)

$$\pi\frac{\Phi^3\sqrt{i_1 i_2}}{\text{Spin}}\dim\Phi^3 i_1\cdot(3+1)\cdot\left\langle \mathbf{D\{2\}}_+^\times\cdot\mathbf{D\{3\}}_+^\times\cdot\mathbf{D\{4\}}_+^\times\right\rangle\cdot$$

$$\cdot\mathbf{InG}_+^\times\cdot\mathbf{Cr}_+^\times\cdot\widehat{E}\breve{E}\cdot\left\langle\mathbf{Chemy}\cdot[\![\circledast\ominus]\!]\cdot\mathbf{\{Genes\}}_+^\times\right\rangle\cdot$$

$$\cdot\odot_+^\times=\frac{\mathbf{65537}}{\Phi^9}\cdot10^{\mathbf{288}}.$$

All this is validated by the such laws of universal harmony as

(542-545) $$\sqrt[e]{\left(\frac{\mathbf{65537}}{\Phi^{9}}\right)^{\Phi\pi i_1}} = \mathbf{9613048};$$

$$\mathbf{9613048}\cdot e^{3\Phi\pi i_1} = \mathbf{2554}\cdot 10^{12};$$

$$\frac{\mathbf{9613048}\cdot E}{10^{12}} - 3 = \cos 72;$$

$$\sqrt[\dim_\Phi \Phi^3 i_1\cdot(3+1)]{\ldots^{\pi\frac{\Phi^3 i_1}{\text{Spin}}}} = \sqrt[5]{\mathbf{55897}\cdot 10^{44}}.$$

What is even more important,

(546) $$\frac{\mathbf{65537}}{\Phi^{9}}\cdot a_e = 1 - \frac{\pi\frac{\Phi^3\sqrt{i_1 i_2}}{\text{Spin}}}{\mathbf{114033}},$$

or else,

(547) $$\frac{1}{\varepsilon}\cdot\frac{\exists}{©@}\cdot\pi\frac{\Phi^3\sqrt{i_1 i_2}}{\text{Spin}}\cdot\mathbf{D\{4\}}_{+}^{\times}\cdot\mathbf{InG}_{+}^{\times} = \frac{10^{65}}{\sin\Theta_W}.$$

To complete topology, it suffices to find out that (548)

$$\frac{\mathbf{65537}}{\Phi^{9}}\times Ghm_e e^{\pm} c\cdot\dim(Ghm_e e^{\pm} c)\times e^{\Phi\pi i_1} = \mathbf{2}\cdot 10^{10}.$$

Therefore, the total picture arrives at (549-551)

$$\sqrt[\chi]{\mathbf{2}\cdot 10^{\mathbf{298}}} = \frac{\Phi}{2}\cdot 10^{59};$$

$$\sqrt[\dim_\Phi \Phi^3 i_1\cdot(3+1)]{\left(\mathbf{2}\cdot 10^{\mathbf{298}}\right)^{\pi\frac{\Phi^3 i_1}{\text{Spin}}}} = \tan 72\cdot 10^{416}.$$

If continued, then (552)

$$\frac{\mathbf{2}\times\mathbf{Ur}_{+}^{\times}\times\{\mathbf{X_M},\mathbf{Y_F}\}_{+}^{\times}\cdot\pi\frac{\Phi^3 i_1}{\text{Spin}}}{10^{63}} - 1 = \frac{\pi\frac{\Phi^3\sqrt{i_1 i_2}}{\text{Spin}}\cdot e^{\Phi\pi i_1}}{10^{8}},$$

*et cetera*, and so on.

But, we should opt for simpler variations and think that the origin of the human soul lies in the Aleph-one which we cannot name what it is exactly. Therefore,

(553) $$\aleph_1^{\Phi^3 i_1} = \mathbf{658361} \cdot 10^{10};$$

(554) $$\left\{\{\mathbf{6,5,8,3,6,1}\}_+^\times \cdot \mathbf{658361}\right\} \cdot \pi \frac{\Phi^3 \sqrt{i_1 i_2}}{\text{Spin}} = e^{\Phi\pi} \cdot 10^{10};$$

(555) $$\left\{\frac{\odot_+^\times \cdot [\![ ⚝ ☺ ]\!] \cdot \mathbf{658361} \times \exp e^{e}}{10^{82}} - 1\right\} 10^6 = \\ = \cosh\cosh\cos i.$$

Note that simply

(556) $$\left\{\frac{\exists}{\text{©@}}\right\}^{\pi \frac{\Phi^3 \sqrt{i_1 i_2}}{\text{Spin}}} \equiv \exp\left\{-e^{e}\right\}.$$

Now the question is: Wether the integer 658361 is really so powerful, or we are going to cheat the uncautious public? False friends occur in this theory as anywhere else. In any event, we are not obliged to be Thomas-Unbelievers towards world harmony, and thus we are in condition to enforce our position in many ways, including

(557) $$\mathbf{658361} \cdot e^{5\Phi\pi e i_1 \sqrt{2}} = \frac{\mathbf{1865921}}{3} \cdot 10^{54}.$$

Yet another logos generated by the same ultra-operator is

(558) $$\frac{e^{5\Phi\pi e i_1 \sqrt{2}}}{\mathbf{658361}} = \sqrt{\Phi\sqrt{i_1 i_2} \cdot 10^{96}}.$$

Now we may or may not go va-banque writing (559)

$$\{\mathbf{658361, 542891047}\}_+^\times = \{\textbf{Soul, Homo}\}_+^\times.$$

If take the risk, then at least (560)

$$\{\textbf{Soul, Homo}\}_+^\times \cdot \pi \frac{\Phi^3 \sqrt{i_1 i_2}}{\text{Spin}} \mathbf{D}\{\mathbf{4}\}_+^\times \cdot \mathbf{InG}_+^\times \div \mathbf{U_E} = \mathbf{66722} \cdot 10^{88}.$$

What is strange, anomalies do compose a configuration:

(561) $$\{\textbf{N, Hi}\}_+^\times \cdot e^{5\Phi\pi e i_1 \sqrt{2}} \cdot \{\textbf{Soul, Homo}\}_+^\times = \Delta_\alpha \aleph_1 \cdot 10^{90}.$$

No less strange is our frame of reference which satisfies

(562) $$\odot_+^\times \cdot \Delta_{Exprm.} = \frac{\mathbf{5202143}}{9} \cdot 10^{41};$$

(563) $$\mathbf{5202143}^{G\Pi} = i_1 i_2 \cdot 10^{24}.$$

And then we come across an unbelievable but exact

(564) $$\sqrt[\otimes]{\left\langle \odot_{+}^{\times} \cdot \Delta_{Exprm.} \aleph_1 \right\rangle^{\oplus}} = \frac{10^{97}}{\cos \Theta_W}.$$

Yet, there is nothing wonderful in here: all this is consistent in the framework of the universal theory we have been desperately trying to relate to the world scientific as well as the lay community. Without deriving eventually the above mystique we were risking to make much ado about nothing. Important is to understand that physics as such becomes at the moment just dogs' game and children's play, as would the nomads put it. The more general concern is that the true theory of mathematical continuum cannot but be a metatheory with all the consequences, including the real possibility to integrate religions and sciences under a single banner.

What else? Sir Isaac Newton dreamed of seeing the divine origin of the Solar system of planets as proved by the rational methods he founded in his **Principia Mathematica**. We much hope that the Sir Newton's immortal soul will have a heavenly joy with this event:

(565) $$\sqrt[\otimes]{\left\{ \odot_{+}^{\times} \cdot \Delta_{Exprm.} \aleph_1 \right\}^{\oplus}} \times \left\{ \pi \frac{\Phi^3 \sqrt{i_1 i_2}}{\text{Spin}} \right\}^{\frac{G}{\Pi}} \equiv 1.$$

The problem of human soul! What is the end destination of man? What is the meaning and purpose of cosmological process?

The unifield theory provides a real chance of approaching this kind of mystical problems. Whether the human civilization survives and lasts longer than the most critical 21st century depends upon the will of united nations. Besides, remind that every single soul will eventually have to pass Purgatorio's test exam, and that the worst place in Dante's Hell is looking for accomodating those, in particular, who had once a great chance but spat upon it.

This work that shortcircuts the Mongolian and Greek past and present had been, in principle, finished in the year

(566) $$\frac{\sqrt[e]{\{\mathbf{Hi,\ N}\}_{+}^{\times \Phi \pi}}}{10^{30}} = \mathbf{2008\ AD}.$$

Yet, it lies unnoticed till the moment in May, 2013. This misfortune compels me to invent an aphorism worth of even Schiller's pen: **The world is not so poor as it seems to itself; instead, the world is just miserable contrary to its own false pretension and self-imagination.**

The Unifield theory is a metatheory, as it should be so, and it should be either accepted in this capacity, or never, with certain consequences in either cases. The metamathematical theorem is

(567) $$\frac{\mathbf{1}}{\mathbf{10}} \mapsto ©@ \cdot \frac{\cos^2 i}{\Phi^3 i_1},$$

where the decimal counting system multifurcates into the real and imaginary realms. One can coin whatever out of it, including a theorem of metaphysical scale and meaning such that (568)

$$\sqrt[\sin\Delta_{Exprm.} \sin\Theta_W \cos\Theta_W \cos 2\Theta_{STR}]{\left\{\frac{\Phi^3 i_1}{©@}\right\}^{\cos^2 i} \cdot \Delta_{Exprm.} \aleph_1} = \frac{\mathbf{858659}}{\mathbf{9}} \cdot 10^{73}$$

and get away safely thanks to the fool-proofness of the universal system of harmony which notably is desparately dreamed of by the species Homo Sapiens since the earliest history.

This tractate we have kicked off with the number, $X$, which does any doubt exist beyond our correct or incorrect will. Of the same independence from our good or bad will is the astronomical configuration of the Solar system of nine planets. But, what we have coined on our own free will is, first, the method of topological densities which, in general, has nevertheless been proved as valid and universal. Secondly, the Solar system topology was configured with respect to the Earth frame of reference. The reason of why so is probably that we, observers, from the Schiller's Mogols to the contemporary sages of some human subspecies, happen to smoke skies over the Earth, nowhere else. Now that we are willing to state the following theorem (492):

(569) $$\sqrt[e]{\left\{\left\{X = \sqrt[\Phi\pi e+1]{\mathbf{10^{90}}}\right\} \odot^{\times}_{+}\right\}^{\pi}} = \cos 2\Theta_{STR} \cdot 10^{59},$$

reducing the problem to the universal energy resonance state.

The number, $X$, embodies and reveals to us what is the mathematical system of harmony in depth. A cosmic dust cloud once lost in the faraway outskirts of the Milky Way galaxy felt free from occasional perturbations of other gravitating masses so that to enjoy the unique chance to be self-organized perfectly in accordance with the laws of harmony. As a result, the unique structure evolved in cosmology and it focused the universal energy flow onto the Blue Globe. Therefore, no wonder that evolution of matter reached its heights giving birth even to the subspecies Homo Sapiens Sapiens, though the picture is going to be a bit spoilt by the political incorrectness of its elite individuals. Thus, the Solar system and the gene code are the two wings of one cosmological process (570):

$$\sqrt[Fluc \cdot Fluc]{\odot^{\times}_{+} \times [\![ 🕸 ☺ ]\!]} = \chi \cdot 10^{23}.$$

Well. How to around up this text correctly? Once physicists in one of major European universities corrected me reminding that there is in physics something called dimension. They didn't know that the Mogol knows something about dimensions since the age 12. Moreover, since my student years I know that the toughest problem in physics is in the end that of dimensions. So, I put

(571) $$\{{}^{\mathbf{Space}}\Phi,\ {}^{\mathbf{Time}}i_1,\ {}^{\mathbf{Mass}}\sqrt{2}\}_{+}^{\times} = \sqrt[\Phi\pi e]{\mathbf{148}\cdot 10^{13}}$$

and see what happens (572--575):

$$\left\{\{\Phi, i_1, \sqrt{2}\}_{+}^{\times} \times h_{Planck}\right\}^{\chi} = 10^{e}\cdot 10^{7};$$

$$\left\{\{\Phi, i_1, \sqrt{2}\}_{+}^{\times} \times h_{Planck}\right\}^{\Phi\pi e i_1} = \Phi^{3} i_1 \cdot 10^{33};$$

$$\Theta^{\chi}_{\textbf{Nuclear Strong Force}} = 10^{e}\cdot 10^{5};$$

$$\mathbf{U}^{G\Pi}_{\mathbf{E}} = 10^{e}\cdot 10^{31},$$

*et cetera*, and so on *ad infinitum*.

To summ up, the Tenggri, or else, the Supreme Intelligence, postulated by Sir Newton, or the Universal Mathematical Machine in our humbler terms is willing to reveal to us some essential features of the final truth. The humankind for ages aspired for the Salvation Messiah-Message and, as it proves to be, not in vain. We relay, though in briefest, the Message received from up above there through the transcosmic communication channels. If Kurt Goedel's metatheorem is true, then such an esoteric event will not be avoidable sooner or later.

Permit me to remind last, but not least, two or three things: *first*, as tells us Samuel Taylor Coleridge, history is accidentality, while science is fatalism; *secondly*, polarities (such as the Mogols and the Hellenes) do shortcircut not only in physics; *thirdly*, opposites such as the Mongols and the Greeks are not contradictory, but complementary, if, of course, believe Niels Bohr's famous principle. So, it will be politically fairly correct a position if accidentality of history and fatalism of science are perceived quietly, without much nostalgia for inquisition. Yes, we do prove Jiordano Bruno's genial theorem, and in due time, not any later. Barbequing that poor monk in Rome was a suicidal risk for civilization; and it is solely Tengri's divine will to correct such a process politically, and in due time, not any later. Not all roads lead to Rome; in critical periods of history roads do, as a rule, lead to Tartaria, - such is my humble apology. The next theorem on the mystery of reality and history I devout to the Bruno's memory (576):

$$\sqrt[\dim_{\Phi}\Phi^{3} i_1\cdot(3+1)]{\{\{\mathbf{2012}-\mathbf{1600}\}\cdot\mathbf{365.256}\}^{\pi\frac{\Phi^{3} i_1}{\text{Spin}}}} = \pi\Phi^{3}\sqrt{i_1 i_2}\cdot 10^{6}\cdot$$

The cosmic space-time is called Tenggri who through His Genius and Wit governs all, including, first and foremost, the human history in the Solar system of planets, a unique phenomenon in the entire infinity of the Universe. You was born in a truly magical, mathematically absolutely harmonious world of ultimate perfection:

(577) $$\sqrt[\frac{\Phi^3\sqrt{i_1 i_2}}{\text{Spin}}]{\left\{\odot_{+}^{\times}\right\}^{\dim_{\Phi}\Phi^3 i_1\cdot(3+1)}} = \mathbf{884078}\cdot 10^{26}$$

which case is commented by

(578) $$\frac{\mathbf{884078}}{\alpha a_e} = \sqrt[\pi]{\mathbf{414192}\cdot 10^{29}};$$

(579) $$\mathbf{414192}^{\exists} = \mathbf{12019}\cdot 10^{28};$$

(580) $$\mathbf{12019}^{\Phi\pi e i_1} \times \frac{\Phi^3\sqrt{i_1 i_2}}{\text{Spin}} = 10^{73}.$$

Natural evolution subject to the laws of harmonization aimed at getting eventually a perfect profduct of extreme complexity and achieved this goal in the Solar system of planets

(581) $$\frac{10^{14}}{\mathbf{884078}\times\left[\!\left[\frac{\text{🕸}}{\text{😐}}\right]\!\right]} + 1 = \cos i.$$

Besides,

(582) $$\sqrt[10]{\arccos \lg\lg\lg\lg\left\{\{Genes\}_{+}^{\times}\times 6\cdot 10^{9}\right\}} = \Phi;$$

(583) $$\sqrt[e]{\left\{\mathbf{542891047}\times 6\cdot 10^{9}\right\}^{2\Phi\pi i_1}} = \Delta_{\beta}\cdot\aleph_1\cdot 10^{83}.$$

Alexander of Macedonia is said to have secluded a nation in the montains of Tartaria. The Mongolian legend relates nearly the same history. But the nomads always re-emerge to the world to make a global noice.

A man asked me whether my calculations are prearranged. I said: Yes, of course. He was very quick to retire. Yes, of course, all this were once prearranged.

## The General Problem of Mass

An ordinary example of the system of mathematical harmony:

(584) $$\mathbf{717937}^{e}\cdot\frac{\mathbf{361933}}{3}=10^{21}.$$

On the other hand,

(585) $$\sqrt[\pi e]{\left\{\pi\frac{\Phi^{3}\sqrt{i_1 i_2}}{\text{Spin}}\cdot 10^{43}\right\}^{\Phi}}=246560952.4,$$

whence we derive a mass (in electronvolts)

(586) $$\sqrt[\pi e]{\left\{\pi\frac{\Phi^{3}\sqrt{i_1 i_2}}{\text{Spin}}\cdot 10^{43}\right\}^{\Phi}}\cdot m_e=125992.6467\,MeV.$$

Now that (not in electronvolts)

(587) $$\sqrt[\pi e]{\left\{\pi\frac{\Phi^{3}\sqrt{i_1 i_2}}{\text{Spin}}\cdot 10^{43}\right\}^{\Phi}}\cdot m_e\dim m\cdot\mathbf{U_E}=\mathbf{717937}\cdot 10^{13};$$

(588) $$m_e\dim m\cdot\mathbf{U_E}=180\Phi\cdot 10^{8}.$$

Given the background energy fluctuation operator of absolute geometry, it is easy to find that (589)

$$\left\{\left\{\sqrt[\pi e]{\left\{\pi\frac{\Phi^{3}\sqrt{i_1 i_2}}{\text{Spin}}\cdot 10^{43}\right\}^{\Phi}}\cdot m_e\dim m\right\}^{Fluc}\right\}^{Fluc}=\mathbf{793}\cdot 19^{25};$$

(590) $$\sqrt[\dim_{\Phi}\Phi^{3}i_1]{\mathbf{793}^{\pi\frac{\Phi^{3}i_1}{\text{Spin}}}}=\frac{10^{19}}{e^{\Phi\pi i_1}}.$$

Gravitation appears to be a very exact phenomenon (591)

$$\sqrt[\otimes]{\left\{\sqrt[\pi e]{\left\{\pi\frac{\Phi^{3}\sqrt{i_1 i_2}}{\text{Spin}}\cdot 10^{43}\right\}^{\Phi}}\cdot m_e\dim m\cdot\frac{G}{\Pi}\right\}^{\oplus}}=\mathbf{4866}\cdot 10^{17}.$$

Subtleties of the origin of mass see in briefest in the following sequence of events (592-593):

$$(125992.6467MeV)\cdot m_e = 64382.24245;$$

$$(125992.6467MeV)\cdot m_e \times \mathbf{D\{2\}}_{+}^{\times}\cdot\mathbf{D\{3\}}_{+}^{\times}\cdot\mathbf{D\{4\}}_{+}^{\times} = \Phi\cdot 10^{82};$$

(594) $$\sqrt[\chi]{\Phi\cdot 10^{82}} = \mathbf{1715905}\cdot 10^{10}.$$

The following is a standard test (595)

$$\lg\lg\lg\lg\left\{\left\{\begin{matrix}\sqrt[\pi e]{\left\{\pi\dfrac{\Phi^3\sqrt{i_1 i_2}}{\text{Spin}}\cdot 10^{43}\right\}^{\Phi}}\cdot\dfrac{1}{1000}\\ \cdot\pi\dfrac{\Phi^3\sqrt{i_1 i_2}}{\text{Spin}}\cdot\dim_{\Phi}\Phi^3 i_1\cdot(3+1)\cdot\\ \cdot\mathbf{D\{2\}}_{+}^{\times}\cdot\mathbf{D\{3\}}_{+}^{\times}\cdot\mathbf{D\{4\}}_{+}^{\times}\end{matrix}\right\}^{\frac{1}{\pi\Pi}}\right\} =$$

$$= \cos\Phi^{10}.$$

If with respect to the fundamental mass spectrum, then

(596) $$\sqrt[\pi e]{\left\{\pi\frac{\Phi^3\sqrt{i_1 i_2}}{\text{Spin}}\cdot 10^{43}\right\}^{\Phi}}\cdot Nm_e\dim m = \frac{10^{13}}{Fluc}.$$

The configuration of boson masses will be (606)

$$\left\{\frac{125992.6467MeV + 80398MeV + 91187.6MeV}{0,511MeV}\right\} = M_B,$$

which can be derived in many ways, including

(597) $$\left\{(M_B)^{Fluc}\right\}^{Fluc} = \frac{\mathbf{76171}}{9}\cdot 10^{13};$$

(598) $$\frac{10^{33}}{(M_B)^{\exists}} = \cos\tilde{\mathbf{i}};$$

(599) $$\exp\left\{\frac{1}{10}\lg\frac{M_B\cdot\mathbf{U}_{\mathbf{E}}}{10^{13}}\right\} = \frac{2}{\Phi}.$$

Just a couple of examples of supersymmetry (600-601)

$$\{\boldsymbol{F},\boldsymbol{B}\}_{+}^{\times 2\Theta_{STR.}} = \frac{\mathbf{29749615}}{3}\cdot 10^{69};$$

$$\frac{\lg\{\boldsymbol{F},\boldsymbol{B}\}_{+}^{\times 2\Theta_{STR.}}}{65537} = a_{e}.$$

What is remarkable,

(602) $$\frac{\{\boldsymbol{F},\boldsymbol{B}\}_{+}^{\times}}{\alpha a_{e}} = \{\boldsymbol{F},\boldsymbol{B}\}_{+}^{\times}\cdot\left(\Delta_{Exprm.}\aleph_{1}\right) = \Phi\cdot 10^{6},$$

and then (603)

$$\left\langle\{\boldsymbol{F},\boldsymbol{B}\}_{+}^{\times}\cdot\{\Delta_{Exprm.},\aleph_{1}\}_{+}^{\times}\right\rangle e^{5\Phi\pi e i_{1}\sqrt{2}}\cdot e^{2\Phi\pi i_{1}} = \frac{\mathbf{18940013}}{3}\cdot 10^{62}.$$

Therefore, we are required to write (604)

$$\sqrt[\pi e]{\left\{\pi\frac{\Phi^{3}\sqrt{i_{1}i_{2}}}{\text{Spin}}\cdot 10^{43}\right\}^{\Phi}}\cdot m_{e}\dim m = \mathrm{M},$$

so that to derive the fermion-boson mass spectrum configuration (605)

$$\mathrm{M}^{F}\times\mathrm{M}^{B} = \frac{\mathbf{57625}}{9}\cdot 10^{34}.$$

This looks something real in terms of universal harmony, but still says nothing particular until after we dig up some more deep hidden subtleties of how the mathematical machinery works. We need to point not again and again that this time we do give priority to pure Gaussian maths making physics quite secondary in our considerations. In the given case of the general problem of mass some more details we are required to learn are

(606) $$\left\{\mathrm{M}^{F}\times\mathrm{M}^{B}\right\}\frac{G}{\Pi} = \frac{10^{42.9999999...}}{\mathbf{12643}};$$

(607) $$\{\mathbf{57625},\mathbf{12643}\}_{+}^{\times}\times\pi\frac{\Phi^{3}i_{1}}{\text{Spin}} \equiv 1;$$

(608) $$\left\{\left\{\mathrm{M}^{F}\times\mathrm{M}^{B}\right\}\{\mathbf{57625},\mathbf{12643}\}_{+}^{\times}\right\}^{\Phi} = \mathbf{U}_{\mathrm{E}}\cdot 10^{74};$$

(609) $$\sqrt[\Phi]{\left\{\left\{\mathrm{M}^{F}\times\mathrm{M}^{B}\right\}\{\mathbf{57625},\mathbf{12643}\}_{+}^{\times}\right\}} = \frac{\mathbf{620335}}{9}\cdot 10^{27}.$$

So far so good. Yet, the nature of the unifield system is such that any worthy loop should be closed, and logically enough. In this respect remind that the very origin of things lies in the spontaneous subquantum leap processes. Therefore, (610)

$$\frac{\mathrm{M}}{\mathbf{620335}} = \frac{10^5}{(©@)^2}(1-\varepsilon),$$

arriving at what we wanted to see on the right. The expression cannot but slightly fluctuate.In this work we make use of an experimental coefficient

$$m_e = 9.10938188,$$

which leads to the fluctuation rate

(611) $$\varepsilon = \frac{1}{7524.681301}.$$

Higher order approximations are always automatical. Leaving intermediary results to the reader's exersize we softland on the self- perturbation effect of the entire geometry as follows (612)

$$\frac{1}{\varepsilon}\cdot\left\langle \pi\frac{\Phi^3\sqrt{i_1 i_2}}{\mathrm{Spin}}\cdot\left\langle \mathbf{D\{4\}}_+^\times \cdot \mathbf{InG}_+^\times\right\rangle\right\rangle\cdot \mathbf{U_E} = \varpi\cdot 10^{74},$$

where

(613) $$\varpi = 1+\Phi^2,$$

is provable as the prototype for angular velocity. Besides, immediately (37)

(614) $$\varpi\varepsilon = \frac{\mathbf{43274}}{9\cdot 10^7}.$$

Apart from absolute geometry there is also algebra of ultra- and super operators. Therefore,

(615) $$\frac{1}{\varepsilon}\cdot e^{5\Phi\pi e i_1\sqrt{2}}\cdot e^{\Phi\pi i_1} = \mathbf{457}\cdot 10^{58}.$$

Now that all physics known up till the moment can be put in a unified form as (39)

(616) $$\sqrt[e]{\left\{\frac{1}{\varepsilon}\cdot\frac{Ghm_e e^{\pm}c\cdot\dim(Ghm_e e^{\pm}c)}{\alpha a_e}\right\}^{2\Phi\pi i_1}} = \frac{\mathbf{539}}{3}\cdot 10^{62}.$$

The three integers previously occured passes the chief test for being not accidental (617):

$$\mathbf{\{x,y,z\}}_+^\times\cdot e^{5\Phi\pi e i_1\sqrt{2}} = \frac{10^{74}}{\mathbf{224311}}.$$

Frankly speaking, we do pursue neither maths, nor physics, but solely fun which is after all the only science. So, we here smell something and calculate the expression and obtain, indeed, an interesting

(618) $$\pi = \mathbf{3.141\,592\,653\,533\,9999...}$$

By the way, a great Mongolian mathematician of the 17th century Sharaid Myanggat told that the only purpose of the exact sciences is to discover ever new methods of approximation of Pi. The Unifield theory is, indeed, merely one of ways of doing so.

In the given instance the mathematical symmetry is broken at the rate

(619) $$\frac{1}{x = 56307719346.77428027920292661620 9}.$$

Does it make sense? Yes, it does much sense, for

(620) $$\boldsymbol{B} = \frac{x \cdot \left\{ \pi \dfrac{\Phi^3 \sqrt{i_1 i_2}}{\text{Spin}} \cdot \dim_\Phi \Phi^3 i_1 \cdot (3+1) \right\} \cdot \cdot \left\{ \mathbf{D\{4\}}_+^\times \cdot \mathbf{InG}_+^\times \right\} \cdot \mathbf{U}_\mathrm{E}}{10^{80.999...}}.$$

closing the loop at the boson sector of matter.

The integer occured above, **18940013**, looks awfully accidental, though incredibly universal. It is actually more than universal successfully passing one of the chief standard tests (621):

$$\sqrt[10]{\arccos \lg\lg\lg\lg \left\{ \frac{1}{\sqrt{2}} \cdot \sqrt[e]{\left\{ \mathbf{18940013} \cdot \pi \frac{\Phi^3 \sqrt{i_1 i_2}}{\text{Spin}} \cdot \cdot \dim_\Phi \Phi^3 i_1 \cdot (3+1) \cdot \cdot \frac{\mathbf{D\{2\}}_+^\times \mathbf{D\{3\}}_+^\times \mathbf{D\{4\}}_+^\times}{\mathbf{InG}_+^\times} \right\}^\pi} \right\}} = \\ = \Phi.$$

But?! What is after all the proper reason for the unit mass? Let us see what follows (622-624):

$$\dagger^{2\chi} = \dagger \cdot 10^{59}; \qquad \dagger = \mathbf{2909681};$$

$$\mathbf{2909681} \cdot \varepsilon = \frac{\mathbf{197049800000}}{9};$$

(625)
$$\left\{\pi\frac{\Phi^3\sqrt{i_1 i_2}}{\text{Spin}}\cdot\dim_\Phi\Phi^3 i_1\cdot(3+1)\right\}\cdot$$
$$\cdot\left\{\frac{\mathbf{D\{2\}}_+^\times\cdot\mathbf{D\{3\}}_+^\times\cdot\mathbf{D\{4\}}_+^\times}{\mathbf{InG}_+^\times}\right\}\times m_e\dim m=\frac{\dagger}{9}\cdot 10^{62}.$$

Whence pure-theoretically (50)

$$m_{Electron}=9.1093818834\,.$$

## The Problem of Angular Velocity

Experimentations and observations in the unifield theory, prompt us to postulate

(626) $$\varpi_{\textit{Angular Velocity Prototype}}=1+\Phi^2.$$

If formally, we will have anything like

(627) $$\frac{1}{\pi}\sqrt[\varpi]{e^{5\Phi\pi i_1 e\sqrt{2}}}=\mathbf{26379}\cdot 10^{10};$$

$$\sqrt[\mathbf{Trig}]{\left\{\varpi^{T_{Cosm.}}\right\}^{Fluc}}=\sqrt{\frac{10^{25}}{180}}.$$

Yet, the genuine proof might be the fact that the Solar system rotations do satisfy an equation

(628) $$\odot_+^\times\cdot e^{\varpi}=\mathbf{15718}\cdot 10^{42},$$

to be played and developed by the reader.

## The Unifield: Fundamental Solutions

Sound instinct needs no
argument; it supllies one.
**Vauvenargues**

In the existing experimental physics are known only seven constants so that to satisfy the following system of mathematical harmony:

$$(629)\qquad \sqrt[e]{\left\{\frac{Ghm_e e^{\pm}c\cdot\dim(Ghm_e e^{\pm}c)}{\alpha a_{Electron}}\right\}^{2\Phi\pi i_1}}=\frac{3}{\mathbf{46189}}\cdot 10^{50};$$

$$(630)\qquad \mathbf{46189}\times\left\{\pi\frac{\Phi^3\sqrt{i_1 i_2}}{\mathrm{Spin}}\cdot\dim_{\Phi}\Phi^3 i_1\cdot(3+1)\right\}\cdot \cdot\mathbf{D\{4\}}^{\times}_{+}\cdot\mathbf{U_E}=\mathbf{81}\cdot 10^{60}.$$

If integrate the whole old and new physics, then a mnemonic

$$(631)\qquad \left\{\begin{array}{c}\dfrac{Ghm_e e^{\pm}c\cdot\dim(Ghm_e e^{\pm}c)}{\alpha a_e}\cdot\\ \cdot\dfrac{\Theta_W\Theta_{STR.}\cdot\{2\times 2\cdot 6\cdot \boldsymbol{FB}\}}{\sin\Delta_{Exprm.}\sin\Theta_W\cos\Theta_W\cos 2\Theta_{STR.}}\end{array}\right\}^{\Phi\pi i_1}=10^{99.999...};$$

$$(632)\qquad \frac{\{...\}^{\Phi\pi i_1}}{1-\sqrt[\pi e]{\dfrac{1}{10^e\cdot 10^{20}}}}=10^{100};$$

$$(633)\qquad \sqrt[\pi e]{10^e\cdot 10^{20}}\cdot\pi\frac{\Phi^3\sqrt{i_1 i_2}}{\mathrm{Spin}}\dim_{\Phi}\Phi^3 i_1\cdot(3+1)=\mathbf{125232}$$

Yet, the fundamental cosmological solution is finally

$$(634)\qquad \left\|\ \frac{\{2\Phi\cdot 10^{17}\}^{\exists}}{\{...\}^{\Phi\pi i_1}}-\sqrt[G\Pi]{\frac{1}{\cos\Theta_W\cdot 10^{10}}}=1\ \right\|$$

Instinctively, the self-perturbation effects of geometry and the instrumental errors in our drawing of algorithmic constructions are the same. So, roughly (635),

$$\left\{\mathbf{125232}\cdot\{\mathbf{3,5,17,257,65537}\}_{+}^{\times}\right\}^{3+1}=\frac{\pi}{2}\cdot 10^{77.9999...},$$

and with more theoreical accuracy (636)

$$\left\{\mathbf{125232}\cdot\{\mathbf{3,5,17,257,65537}\}_{+}^{\times}\right\}^{\Phi\pi}=\frac{\mathbf{706}}{3}\cdot 10^{97.0000000...}.$$

Therefore, conceptually, aphoristically and mnemonically

(637) $$\sqrt[\Phi\pi i_1]{\mathbf{10^{100}}}=\text{Unifield}\cdot$$

Now we complete the bare unifield with atomic physics (638):

$$\left\|\sqrt[e]{\left\{\begin{array}{c}\dfrac{2N\cdot Ghm_e e^{\pm}c\cdot\dim(Ghm_e e^{\pm}c)}{\alpha a_e}\cdot\\ \cdot\dfrac{\Theta_W\Theta_{STR.}\cdot\{2\times 2\cdot\mathbf{6}\cdot\boldsymbol{FB}\}\times\mathbf{92}}{\sin\Delta_{Exprm.}\sin\Theta_W\cos\Theta_W\cos 2\Theta_{STR.}}\end{array}\right\}^{\pi}}=-\frac{10^{23.9999...}}{\cos\Phi^{10}}\cdot\right\|$$

Know that experimental values can never be better than we quote in this work. Self-perturbations of geometry defined by the chaotic turbulence of all the competing constants, operators and configurations do never vanish. Therefore, the final value of the previous unifield configuration will be determined by the following factors: 1. Logical inevitability; 2. Formal mathematical sufficiency; 3. The cosmlogical average of turbulence effects of geometry.

In the given case the self-perturbative effect of the entire geometry

$$\mathbf{Turb}=25426.69522$$

is defined in as many different ways as the reader could find out applying only to the innate mechanisms of absolute geometry. For example,

(639) $$\frac{137000}{\mathbf{Turb}}\equiv\Phi^3 i_1;$$

$$\mathbf{Turb}\cdot\widehat{E}\breve{E}\equiv\frac{1}{\cos i};$$

$$\mathbf{Turb}\cdot\left\{\widehat{E}\breve{E}\cdot\mathbf{U_E}\cdot E\right\}\equiv e^{\Phi};$$

$$\mathbf{Turb}^{\pi}\equiv\tan 66\tan 72;$$

(643) $$\mathbf{Turb}^2 \equiv \Phi \pi i_1;$$

(644) $$\frac{\mathbf{Turb}}{\mathbf{360}} \equiv \frac{10}{\pi - 3},$$

*et cetera ad infinitum.*

In each case of the above we will have exact pictures such as

(645) $$\frac{\mathbf{Turb}^{\Phi \pi e i_1}}{\mathbf{D}\{\mathbf{2}\}_+^\times \cdot \mathbf{D}\{\mathbf{3}\}_+^\times \cdot \mathbf{D}\{\mathbf{4}\}_+^\times} = 1 + \sqrt[G11]{\frac{10}{170029}}.$$

## The Cosmic Unifield
## &
## the Human Mind

Now that with the absolute inevitability of pure mathematics we stumble over the mother of all problems. This is the problem of **the relationship between the Cosmos and the human mind**. Mind bifurcates into logic and psyche. Logically, you cannot but conceive all the beauty of the Cosmos as defined above in mathematical cosmology. Yet, your psyche will readily find out a dozen or more reasons to oppose. Such is the human nature, a dreadful phenomenon in Nature. We are born with quite a shrewd intelligence because we do evolutionally succeed to a quite coward and agressive species of apes, as it is testified by the latest data in anthropology. I advise to the young generations only one strategy: Always know where is your brilliant intelligence and where is your coward psyche.

Whether human mind makes a component of the cosmic unifield? Whether the state of the Universe is independent from human thought? Is it of absolutely no concern to the Cosmos what you think, say, of me at the moment? This is the final question asked for centuries by the sciences and religions alike. You now know that the cosmic unifield is eventually determined by perturbations. Whether human thought, our psychological state can be a perturbative effect, too? We have found that (646)

$$\frac{\left\{2\Phi \cdot 10^{17}\right\}^{\exists}}{\left\{\frac{2N \cdot Ghm_e e^{\pm} c \cdot \dim(Ghm_e e^{\pm} c)}{\alpha a_e} \cdot \frac{\Theta_W \Theta_{STR.} \cdot \{2 \times 2 \cdot \mathbf{6} \cdot \boldsymbol{FB}\} \times \mathbf{92}}{\sin \Delta_{Exprm.} \sin \Theta_W \cos \Theta_W \cos 2\Theta_{STR.}}\right\}} - 1 = \frac{1}{\mathbf{Turb}}.$$

Now are left just see (647)

$$\mathbf{Turb}\times\left\{\left[\left[\frac{\text{🕸}}{\text{😐}}\right]\right]\cdot(\mathbf{Cr}_{+}^{\times}\cdot\mathbf{22727})\right\}\mathbf{542891047}=\frac{10^{\mathbf{46.0000...}}}{\pi\dfrac{\Phi^{3}i_{1}}{\text{Spin}}}.$$

Consequently, the human existence in the cosmic space-time is a full blown perturbative factor in its own right.

At this point I could in principle and in quiet conscious around up with this book believing that the reader will always think good about the Universe and in particular of the Solar system where man is the inverse operator and *vice versa*:

(648) $$\odot_{+}^{\times}\cdot\left\{\widehat{E}\breve{E}\times\left\{\left\{\frac{\dim E}{2}\right\}^{2}\right\}^{2}\right\}=\frac{10^{76.0000...}}{\mathbf{542891047}},$$

where the symmetry is broken by the Stonehenge code of geometry

(649) $$\sqrt[e]{y^{2\Phi\pi i_{1}}}=\Phi(©@)^{2}\cdot 10^{56}$$

## Monochord-Ultrasting

Without adventure civilization is in full decay.
**Alfred North Whitehead**

Pythagoras thought that his monochord will have sufficed to meet any requirements for analysis and synthesis. Indeed, we have seen that

$$\Phi\left\{\{\mathbf{X_{M}},\mathbf{Y_{F}}\}_{+}^{\times}\right\}^{\Phi}=\frac{\mathbf{828371}}{3}\cdot 10^{62}.$$

The total superstructure-ultrastring topology shall be (650)

$$\frac{\pi\dfrac{\Phi^{3}\sqrt{i_{1}i_{2}}}{\text{Spin}}\dim_{\Phi}\Phi^{3}i_{1}(3+1)\dfrac{\mathbf{D\{2\}_{+}^{\times}\cdot D\{3\}_{+}^{\times}\cdot D\{4\}_{+}^{\times}}}{\mathbf{InG_{+}^{\times}}}\{\mathbf{X_{M}},\mathbf{Y_{F}}\}_{+}^{\times}}{1-\sqrt[\chi]{\dfrac{1}{\exists}}}=$$

$$=\cos\Theta_{W}\cdot 10^{\mathbf{108}};$$

(651) $$\mathbf{108=1^{1}2^{2}3^{3}}.$$

## The System of Absolute Operators

Operators are many, though we refer only to frequently used ones. Previously we have found out an operator being a multidigit integer. Thus,

(652) $$\{\oplus\otimes\}\{\Phi i_1\}\{\chi\cdot G\Pi\}\cdot\mathbf{2909681}=\frac{\mathbf{41884900000}}{9};$$

$$\{\oplus\otimes\}\{\Phi i_1\}\{\chi\cdot G\Pi\}\cdot\mathbf{2909681}\times$$
$$\times e^{5\Phi\pi i_1 e\sqrt{2}}\cdot e^{\Phi\pi i_1}=\mathbf{\{3,5,17,257,65537\}}_+^\times\cdot 10^{52},$$

$$\pi e\{\oplus\otimes\}\{\Phi i_1\}\{\chi\cdot G\Pi\}\cdot\mathbf{2909681}\times$$
$$\times e^{5\Phi\pi i_1 e\sqrt{2}}\cdot\frac{\mathbf{239704}}{9}=10^{69};$$

$$\pi e\{\oplus\otimes\}\{\Phi i_1\}\{\chi\cdot G\Pi\}\cdot\mathbf{2909681}\times$$
$$\times e^{5\Phi\pi i_1 e\sqrt{2}}\cdot\frac{\exists}{©@}=\frac{\mathbf{28426358}}{3}\cdot 10^{59};$$

$$\sqrt[5]{\pi e\{\oplus\otimes\}\{\Phi i_1\}\{\chi\cdot G\Pi\}\cdot\mathbf{2909681}\times e^{5\Phi\pi i_1 e\sqrt{2}}\cdot\frac{\exists}{©@}\times\left\{\frac{Ghm_e e^{\pm}c\cdot\dim(Ghm_e e^{\pm}c)}{\alpha a_e}\right\}}=\frac{10^{20}}{\mathbf{75634}};$$

$$\mathbf{75634}^{\Phi}=\frac{9}{\mathbf{114898}}\cdot 10^{12}.$$

$$\odot_+^\times\cdot e^{\varpi}\cdot\frac{G}{\Pi}\left\{1+\sqrt[G\Pi]{\frac{a_e}{\mathbf{10^{12.9999...}}}}\right\}=\frac{\pi}{\Phi}\cdot 10^{47};$$

$$\frac{\{Genes\}_+^\times}{\{\oplus\otimes\}\{\Phi i_1\}\{\chi\cdot G\Pi\}\cdot\mathbf{2909681}}=\mathbf{\{X_M,Y_F\}}_+^\times\cdot 10^{24}.$$

$$\frac{\{Genes\}_{+}^{\times}\cdot\widehat{E}\breve{E}}{e^{5\Phi\pi i_1 e\sqrt{2}}}=\sin\Theta_W\cdot 10^{44.0000...}.$$

$$\mathbf{\{5,4,2,8,9,1,4,7\}_{+}^{\times}\cdot 542891047}\equiv\frac{1}{\exists}:$$

$$\frac{\mathbf{D\{4\}_{+}^{\times}}}{\mathbf{\{5,4,2,8,9,1,4,7\}_{+}^{\times}\cdot 542891047}}=\Phi\cdot 10^{30.0000...},$$

$$\odot_{+}^{\times}\cdot\mathbf{542891047}\cdot\widehat{E}\breve{E}=\frac{10^{76.0000...}}{\left\{\left\{\frac{\dim E}{2}\right\}^2\right\}^2};$$

$$\{\mathbf{U_E}\}^{Fluc\cdot Fluc\cdot Fluc\cdot Fluc}=\mathbf{4396}\cdot 10^{77}.$$

$$\left\langle\left\langle\left\{\left\{\frac{\dim E}{2}\right\}^2\right\}^2\right\rangle^2\right\rangle^2=\frac{1}{100a_e}=\frac{\aleph_1}{100}\cdot$$

$$\left\{\pi\frac{\Phi^3\sqrt{i_1 i_2}}{\mathrm{Spin}}\cdot\dim_\Phi\Phi^3 i_1\cdot(3+1)\right\}\cdot\mathbf{D\{2\}_{+}^{\times}D\{3\}_{+}^{\times}D\{4\}_{+}^{\times}}\times$$

$$\times Ghm_e e^{\pm}c\cdot\dim(Ghm_e e^{\pm}c)=\mathbf{24828}\cdot 10^{80.0000000...};$$

$$\frac{\mathbf{24828}}{\alpha a_e}=\sqrt[\pi e]{\frac{10^{81}}{\sqrt{2}}},$$

Is is a fun that anomalies are designed to compose

$$\frac{\mathbf{\{1292049,310952\}_{+}^{\times}}}{\mathbf{542891047}}=10000\left\{\Delta_1\cdot\frac{65537}{\Phi^9}\right\}\left\{1+\frac{1}{1000\Phi}\right\}.$$

Formally,

(667) $$\left\{\Delta_1\cdot\frac{65537}{\Phi^9}\right\}\cdot\alpha a_e=1+\sqrt[\chi]{\sqrt[\chi]{\frac{1}{\Phi^3 i_1\cdot 10^{63}}}}.$$

## Exersizes in Computing the Reality

(668) $$\mathrm{A} = \left\{ \frac{Ghm_e e^{\pm} c \cdot \dim(Ghm_e e^{\pm} c)}{\alpha a_e} \right\};$$

$$\mathrm{B} = \left\{ \pi \frac{\Phi^3 \sqrt{i_1 i_2}}{\mathrm{Spin}} \cdot \dim_{\Phi} \Phi^3 i_1 \cdot (3+1) \right\};$$

$$\Gamma = \frac{\mathbf{D\{2\}}_+^{\times} \mathbf{D\{3\}}_+^{\times} \mathbf{D\{4\}}_+^{\times}}{\mathbf{InG}_+^{\times}}.$$

$$\mathrm{AB\Gamma} \cdot \tilde{E}\breve{E} = \frac{\mathbf{3}}{\mathbf{11}} \cdot 10^{99};$$

$$\sqrt[3]{\mathrm{AB\Gamma} \cdot \tilde{E}\breve{E}} = \mathbf{6485} \cdot 10^{29};$$

$$\frac{\sqrt[Fluc \cdot Fluc \cdot Fluc]{\mathrm{AB\Gamma} \cdot \tilde{E}\breve{E}}}{\mathbf{2380486} \cdot 10^{13}} = \cos i.$$

$$\ln \lg \frac{\mathbf{2380486} \cdot \mathbf{U_E}}{\Phi^3 i_1} = \boldsymbol{B};$$

$$\mathbf{2380486} = \sqrt[\chi]{\frac{10^{31}}{(©@)^2}}.$$

$$\left\{ \mathrm{AB\Gamma} \cdot \tilde{E}\breve{E} \times \boldsymbol{FB} \right\} \cdot e^{5\Phi\pi i_1 e\sqrt{2}} = \mathbf{886677} \cdot 10^{147}.$$

$$\mathbf{886677}^{\exists} = \frac{\mathbf{27851}}{3} \cdot 10^{30}.$$

$$\frac{\mathrm{AB\Gamma} \cdot \tilde{E}\breve{E} \times \boldsymbol{FB}}{e^{5\Phi\pi i_1 e\sqrt{2}}} \left\{ 2 \times (\mathbf{6}_{Leptons} + \mathbf{6}_{Quarks}) \right\} = \mathbf{23843} \cdot 10^{42};$$

$$\mathbf{23843}^{\Phi\pi e i_1} = \mathbf{866} \cdot 10^{73.999999...} = \mathrm{Spin} \cdot 10^{76.9999...}.$$

Particles interact in

$$12\times 12$$

ways. Therefore,

$$\sqrt[e]{\{\mathbf{23843}\{12\times 12\}\}^{\Phi\pi}}=\frac{\mathbf{5}}{\mathbf{3}}\cdot 10^{12};$$

$$\frac{\{\mathbf{3,5,17,257,65537}\}_{+}^{\times}}{\mathbf{2384300000}}=\frac{\Delta_1}{a_e};$$

$$\frac{\text{🕸}}{\mathbf{23843}}=\frac{\mathbf{1673989}}{9}10^{7};$$

$$\frac{\text{😐}}{\mathbf{167398900}}=i_1.$$

$$\mathbf{Uni}=\Delta_1-\mathbf{137}.$$

$$\mathbf{23843}\times\text{😐}=\mathbf{Uni}\cdot\mathbf{10}^{\mathbf{12.99991919}}.$$

$$\frac{\mathbf{23843}\times\text{😐}}{\mathbf{Uni}}=\mathbf{360}\cdot\{\mathbf{3,5,17,257,65537}\}_{+}^{\times}\cdot\Phi\pi i\left\{1+\frac{1}{z}\right\},$$

$$z=17032.23898,$$

$$\sqrt[e]{z^{2\Phi\pi i_1}}=10^{e}\cdot 10^{52}.$$

$$\{Genes\}_{+}^{\times}\times\mathbf{Uni}=\frac{10^{78}}{\mathbf{1188}}$$

$$\frac{\{Genes\}_{+}^{\times}}{[\![\text{🕸😐}]\!]}=\sqrt[\mathbf{Uni}]{\sqrt[\mathbf{Uni}]{\mathbf{9}\cdot 10^{12}}}.$$

(28) $$\sqrt[10]{\arccos\lg\lg\lg\lg\frac{\sqrt[e]{\{\odot_{+}^{\times}\cdot\mathbf{9}\}^{\Phi\pi}}}{\left\{\left\{\frac{\dim E}{2}\right\}^{2}\right\}^{2}}=\Phi}.$$

Thirdly, cosmology has achieved from time Zero up till the moment the following planetary life configuration as well as our ability to write it:

$$\sqrt{\left\{\left\{\odot^{\times}_{+}\frac{\{Genes\}^{\times}_{+}}{[\![\text{🕸}\,\text{😐}]\!]}\right\}^{\mathbf{Uni}}\right\}^{\mathbf{Uni}}} \equiv \pi\cos\frac{180}{\pi}\cdot 10^{12}.$$

It is not excluded that the Greeks meant that, notably,

$$\frac{\sqrt{\left\{\left\{\odot^{\times}_{+}\frac{\{Genes\}^{\times}_{+}}{[\![\text{🕸}\,\text{😐}]\!]}\right\}^{\mathbf{Uni}}\right\}^{\mathbf{Uni}}}}{1+\sqrt[\dim_{\Phi}\Phi^{3}i_{1}\cdot(3+1)]{\left\{\frac{1}{\mathbf{310952}}\right\}^{\pi\frac{\Phi^{3}i_{1}}{\text{Spin}}}}} = \pi\cos\frac{180}{\pi}\cdot 10^{12}.$$

Any given structure in the Universe is manufactured by the physical forces. In this respect the species Homo Sapiens in the Solar system looks equivalently either

$$\sqrt[\Phi\pi i_{1}e]{\left\{\odot^{\times}_{+}\frac{\{Genes\}^{\times}_{+}}{[\![\text{🕸}\,\text{😐}]\!]}\right\}\cdot\left\{\Delta_{Exprm.}\Theta_{W}\Theta_{STR.}\right\}} = \frac{\mathbf{1459000}}{3},$$

or

(695) $$\sqrt[\Phi\pi i_{1}e]{\left\{\odot^{\times}_{+}\frac{\{Genes\}^{\times}_{+}}{[\![\text{🕸}\,\text{😐}]\!]}\right\}\cdot\left\{\Delta_{Exprm.}\Theta_{W}\Theta_{STR.}\right\}}\times\Phi\frac{\Phi\pi}{4} = 10^{6}.$$

## Functions of Forces

> It is awkward to carry uncoined gold: that's what a thinker does who has no formulae.
>
> **Friedrich Nietzsche**

The fermion-boson interactions are subject to a general formula

(696) $$\left\{\Delta_{Exprm.}\Theta_{W}\Theta_{STR.}\right\}\cdot\mathbf{U}_{\mathbf{E}} = \sqrt[\pi]{\mathbf{230696}\cdot 10^{40}}.$$

In the end the problem is that of energy and entropy

(697) $$\left\{\Delta_{Exprm.}\Theta_{W}\Theta_{STR.}\right\}^{FB} = \frac{\mathbf{28487234}}{9}\cdot 10^{11};$$

(698) $$\mathbf{28487234}\cdot \overset{\frown}{E}\breve{E} = \mathbf{72605}\cdot 10^{25}\cdot$$

Naturally,

(699) $$\left\{\Delta_{Exprm.}\Theta_{W}\Theta_{STR.}\right\}^{FB\cdot FB} \equiv \cosh\cosh\cosh\cos i\,.$$

It is important to see again and again the Solar system as a megascale crystal

(670) $$\left\{\odot_{+}^{\times}\cdot \mathbf{Cr}_{+}^{\times}\right\}\left\{\Delta_{Exprm.}\Theta_{W}\Theta_{STR.}\right\} = \mathbf{402}\cdot 10^{66},$$

which notably makes possible

(671) $$\left\{\Delta_{Exprm.}\Theta_{W}\Theta_{STR.}\right\}\cdot\left\{\circledast\cdot \mathbf{Cr}_{+}^{\times}\right\} = \mathbf{42266}\cdot 10^{36};$$

(672) $$\left\{\Delta_{Exprm.}\Theta_{W}\Theta_{STR.}\right\}\cdot\left\{[\![\circledast\odot]\!]\cdot \mathbf{Cr}_{+}^{\times}\right\} = \mathbf{9}\cdot 10^{49}.$$

The general law of evolution of matter structures is

(673) $$\left\{\Delta_{Exprm.}\Theta_{W}\Theta_{STR.}\right\}^{T_{Cosm.}\cdot Fluc}\cdot \mathbf{U_E} = \frac{10^{23}}{\mathbf{360}}.$$

A detailed view of the evolutional process is

(674) $$\sqrt[e]{\left\{\left\{\Delta_{Exprm.}\Theta_{W}\Theta_{STR.}\right\}^{FB\cdot T_{Cosm.}\cdot Fluc}\right\}^{\Phi\pi}} = \mathbf{2316052}\cdot 10^{65};$$

(675) $$\mathbf{2316052}\cdot \mathbf{U_E} = \dim^2 E\cdot 10^{15} = \frac{\mathbf{Cr}_{+}^{\times}\cdot 2\Theta_{STR.}}{10^{5}}.$$

What here surprises is the theoretical precision of interaction mechanics. This implies, pragmatically, that the Theta parameters of nuclear forces have been defined right. What is vital, from now onward theories will have priority before experiment, as the latter is understood now.

The total of fine tuning mechanisms comes to

(676) $$\left\{\Delta_{Exprm.}\Theta_{W}\Theta_{STR.}\right\}\cdot FB\cdot T_{Cosm.}\cdot Fluc = \sqrt[\Phi\pi e]{\frac{\mathbf{690406}}{3}\cdot 10^{77}}$$

with the consequence that

(677) $$\mathbf{690406}\cdot \odot_{+}^{\times} = 180\Phi\cdot 10^{48};$$

(678) $$\mathbf{690406}\cdot \mathbf{Cr}_{+}^{\times} = \frac{\mathbf{162181}}{3}\cdot 10^{20}.$$

The fact that the Solar system astronomical configuration is of absolute perfection is seen in

(679) $$180\Phi\cdot \mathbf{162181} = \frac{\mathbf{141703760}}{3}.$$

Whence, foreseeably,

(680) $$\mathbf{141703760}\cdot[\![\text{🕸}]\!]\cdot\mathbf{Cr}_{+}^{\times}\cdot\mathbf{22727}=\frac{10^{48}}{\cos\Theta_W};$$

(681) $$\frac{\mathbf{141703760}}{3}\cdot\left[\!\!\left[\frac{\text{🕸}}{\text{😐}}\right]\!\!\right]\cdot\mathbf{Cr}_{+}^{\times}\cdot\mathbf{22727}=\exists.$$

Before we have had the unifield such that

(682) $$\frac{Ghm_e e^{\pm}c\cdot\dim(Ghm_e e^{\pm}c)}{\alpha a_e}\cdot\frac{\Theta_W\Theta_{STR.}\cdot\{2\times2\cdot6\cdot\boldsymbol{FB}\}}{\sin\Delta_{Exprm.}\sin\Theta_W\cos\Theta_W\cos2\Theta_{STR.}}=\text{Unifield}$$

Its shorter form is an exact

(683) $$\frac{\cdots}{\sin\Delta_{Exprm.}}=\mathbf{4589595}\cdot10^{8}.$$

So, the two ends of geometry meet in

(684) $$\text{Unifield}\times180\Phi\cdot162181\cdot141703760=\pi\frac{\Phi^3\sqrt{i_1 i_2}}{\text{Spin}}\cdot10^{30}.$$

Yet, ultimately (685),

$$\left\{{}^{1+\Phi\pi e}\!\!\sqrt{\mathbf{10^{90}}}\cdot e^{5\Phi\pi i_1 e\sqrt{2}}\right\}^{\sin\Delta_{Exprm.}\sin\Theta_W\cos\Theta_W\cos2\Theta_{STR.}}=\frac{\mathbf{24257314}}{9},$$

whence finally

$$\Delta_{\textit{Theoretical \& Experimental}}=137.035999000\ldots$$

## The Unifield Pro & Contra 'Black Holes'

Cosmology begins with the notion of space-time. In the unified theory the 4-Dim quantum-gravitational space-time turns out to be

(686) $$\left\{ \begin{array}{l} \pi \dfrac{\Phi^3 \sqrt{i_1 i_2}}{\text{Spin}} \cdot \dim_\Phi \Phi^3 i_1 \cdot (3+1) \cdot \\ \cdot \left\{ \begin{array}{l} [G = 6.673] \cdot \\ \cdot [h = 6.62606876000...] \end{array} \right\} \end{array} \right\}^{\Phi \pi e} = \frac{\mathbf{7837048}}{3} \cdot 10^{50};$$

(687) $$\sqrt[e]{\mathbf{7837048} \cdot \left\{ Gh \cdot \Phi^3 i_1 \right\}^{2\Phi\pi i_1}} = \frac{\mathbf{115}}{9} \cdot 10^{43}.$$

We may continue and obtain a sequence of constants

(688) $$\frac{\left\{ \left\{ \dfrac{\exists}{©@} \left\{ \dfrac{Ghm_e e^{\pm} c \cdot \dim(Ghm_e e^{\pm} c)}{\alpha a_e} \right\} \Theta_W \Theta_{STR.} \right\} \boldsymbol{FB} \right\}}{\mathbf{115}} =$$

$$= \frac{9}{\mathbf{314669}} \cdot 10^{17};$$

(689) $$\mathbf{314669} \cdot e^{5\Phi\pi e i_1 \sqrt{2}} = \frac{\mathbf{26754964}}{9} \cdot 10^{53}.$$

A much shorter way to describe space-time is

(690) $$\sqrt[e]{\left\{ \pi \frac{\Phi^3 \sqrt{i_1 i_2}}{\text{Spin}} \cdot Gh \right\}^{\Phi \pi i_1}} = \mathbf{9679799};$$

(691) $$\mathbf{9679799}^{\Phi\Phi} = \frac{\mathbf{5839246}}{3} \cdot 10^{12}.$$

What is remarkable,

(692) $$\mathbf{7837048}^h = \sqrt[\Phi]{\mathbf{8196315} \cdot 10^{67}}.$$

And also

(693) $$\mathbf{7837048} \times \chi \cdot G\Pi = \mathbf{143051993};$$

(694) $$\sqrt{\frac{\mathbf{7837048}}{\Phi\pi e i_1 \cdot \oplus \otimes \cdot \chi \cdot G\Pi \times e^{\Phi\pi i_1}}} \cdot 10^{54} = e^{5\Phi\pi e i_1 \sqrt{2}}.$$

It is unbelievable that

(695) $$\mathbf{7837048} \times \odot_{+}^{\times} = \frac{\mathbf{9916967}}{3} \cdot 10^{45};$$

(696) $$\lg \frac{\mathbf{9916967}^{\Phi\Phi\Phi\Phi\Phi}}{39} \cdot \frac{1}{65537} = a_e = \aleph_1.$$

In physics it is thought that collapsed masses become the so called ‘black holes’ remaining, however, fairly hypothetical odjects. The Bekenstein-Hawking entropy makes them physically more real from the theoretical point of view. This entropy is usually written as

(697) $$S_{BH} = A\frac{c^3}{4Gh}.$$

The main problem here is the surface area $A$ which can be in our terms put as

(698) $$A = \Phi^2;$$

(699) $$A = 4\pi\Phi^2.$$

But, in the existing physics it is usually

(700) $$A = 16\pi\left\{\frac{GM}{c^2}\right\}^2.$$

We may imagine first that a nucleon is a natural black hole

(701) $$S_{BH} = 16\pi\left\{\frac{G \dim G \cdot Nm_e \dim m}{(ci_1)^2}\right\}^2 \cdot \frac{(ci_1)^3}{4G \dim G \cdot h \dim h}.$$

Then, quite unexpectedly,

(702) $$\mathbf{404319500} = \left\{\frac{G}{\Pi} \cdot \frac{Nm_e \dim m}{(ci_1)^2}\right\}^2;$$

$$\mathbf{404319500}^{\Phi\pi i_1} = \sin\Theta_W \cdot 10^{56.00000\ldots}.$$

But now quite foreseeably,

(703) $$S_{BH} = 10000\Delta_{Exprm.}\aleph_1.$$

And, of course,

(704) $$\sqrt[e]{\{S_{BH}\}^{\Phi\pi i_1}} = \exp\exp e \cdot 10^{15};$$

(705) $$\{S_{BH}\}^{\chi} = \mathbf{878} \cdot 10^{43}.$$

Yet, a surprisingly exact

(706) $$S_{BH} \cdot e^{5\Phi\pi i_1 e\sqrt{2}} = \frac{10^{69}}{\mathbf{895051}},$$

justified by

(707) $$\mathbf{895051}^{\Phi\pi e} = \frac{9}{\mathbf{5177359}} \cdot 10^{88}.$$

Now we move to the Hawking black hole temperature

(708) $$T_H = \frac{\hbar c^3}{8\pi G M k_B} = \frac{\hbar \dim \hbar (c i_1)^3}{8\pi G \dim G \cdot N m_e \dim m \cdot k_B};$$

(709) $$T_H = \frac{1}{59607.54919...}.$$

It is already not surprising that

(708) $$\frac{\{\widehat{E}\breve{E} \cdot \mathbf{U_E} \cdot E\}}{T_H} = \Delta_{Exprm.} \aleph_1 \cdot 10^{37}.$$

Naturally,

(709) $$\frac{S_{BH}}{T_H} = \frac{\mathbf{2114789}}{3} \cdot 10^8;$$

(710) $$\mathbf{2114789} \cdot e^{5\Phi\pi i_1 e\sqrt{2}} = \mathbf{2} \cdot 10^{60}.$$

Now we definitely know what will be the final result:

(711) $$\mathbf{2114789} \cdot \frac{\left\{\dfrac{G h m_e e^{\pm} c \cdot \dim(G h m_e e^{\pm} c)}{\alpha a_e}\right\} \cdot \Theta_W \Theta_{STR.} \cdot \{12 \times 2\} \cdot \boldsymbol{FB}}{\sin \Delta_{Exprm.} \sin \Theta_W \cos \Theta_W \cos 2\Theta_{STR.}} \cdot \mathbf{D\{4\}}_{+}^{\times} = \exists.$$

Or equivalently,

(712) $$\frac{S_{BH}}{T_H} \cdot [...] \equiv \Phi\sqrt{i_1 i_2}.$$

The power of integral operator conceiled in the womb of Nature

(713) $$\mathbf{2114789} \frac{2\Theta_{STR.}}{\cos 2\Theta_{STR.}} = \mathbf{360} \cdot 10^6.$$

With respect to the standard global topology we have

(714)
$$\mathbf{2114789}\cdot\pi\frac{\Phi^{3}\sqrt{i_1 i_2}}{\text{Spin}}\dim_{\Phi}\Phi^{3}i_1(3+1)\cdot$$
$$\cdot\frac{\mathbf{D\{2\}_{+}^{\times}\cdot D\{3\}_{+}^{\times}\cdot D\{4\}_{+}^{\times}}}{\mathbf{InG_{+}^{\times}}}\mathbf{U_E}=\mathbf{12}\cdot 10^{81}.$$

Now that cosmobiology owes to the following system of greatest exactitude:

(715)
$$\llbracket \text{🕸} \rrbracket\cdot S_{BH}=\frac{\mathbf{47201719}}{9}\cdot 10^{19};$$

$$\mathbf{Cr_{+}^{\times}}\cdot\llbracket \text{🕸} \rrbracket\cdot S_{BH}=\frac{\mathbf{1232}}{3}\cdot 10^{42};$$

$$\left\{\mathbf{Cr_{+}^{\times}}\cdot\mathbf{22727}\right\}\cdot\llbracket \text{🕸} \rrbracket\cdot S_{BH}=\frac{\mathbf{28}}{3}\cdot 10^{48};$$

$$\left\{\llbracket ☹ \rrbracket\left\{\Phi\cdot\Delta_{Exprm.}\cdot i_1\right\}\right\}\cdot\frac{S_{BH}}{T_H}=\frac{\mathbf{127}}{3}\cdot 10^{23};$$

(719)
$$\frac{\odot_{+}^{\times}}{S_{BH}}=\frac{\mathbf{107}}{3}\cdot 10^{34}.$$

To spare space, we leave to the reader the logic and philosophy of 'black holes' to be derived on his/her own exempt making the only remark about the following bifurcation process inside the mathematical continuum:

(720)
$$\sqrt[\Phi\pi e+1]{10^{90}} \Leftrightarrow \text{Cosmology}$$
$$\swarrow \quad \searrow$$
$$\Delta_{Exprm.} \qquad \aleph_1$$
$$\swarrow \quad \searrow$$
$$S_{BH} \qquad T_H$$
$$\swarrow \qquad\qquad \searrow$$
Physics Biology

Yet, finally, by definition

(721) $$4\pi\Phi^2 \cdot S_{BH} = \sqrt[\pi]{\frac{10^{34}}{\Phi^3 i_1}},$$

which is the nucleon black hole.

## Planck Density

In terms of absolute calculus the Planck density is by definition

(722) $$\rho_{Planck} = \frac{c^5}{\hbar G^2} \cdot \frac{\sqrt{2}}{\Phi^3} \equiv \mathbf{InG}_{+}^{\times}.$$

Note that

(723) $$\left\{10000 \frac{\mathbf{InG}_{+}^{\times}}{\alpha a_e}\right\}^{\pi} = \{\dim E\}^2 \cdot 10^{70};$$

(724) $$\mathbf{Cr}_{+}^{\times} \cdot 2\Theta_{STR.} = \{\dim E\}^2 \cdot 10^{20}.$$

The Planck density in nondimensional form and with Dirac's constant comes, by definition, to

(725) $$\rho_{Planck} = \frac{c^5}{\hbar G^2} \equiv \frac{10^{100}}{360 \cdot \Phi^3 i_1},$$

where we derive the Euclidean and post-Euclidean volumes of geometry.

Accurately,

(726) $$\rho_{Planck} = \frac{c^5}{\hbar G^2} = \frac{10^{100}}{360 \cdot \Phi^3 i_1}\left\{1 + \frac{1}{Z}\right\};$$

(727) $$Z = 3006.3171960...;$$

(728) $$Z^{\pi} \cdot \frac{9}{\mathbf{65537}} = a_e \cdot 10^{10}.$$

The absolute geometry is precise intuitively, logically and calculationally. If accurately, we have the light speed as

(729) $$\frac{\mathbf{299792458}^{\pi}}{\mathbf{427285} \cdot 10^{21}} = 1 - \frac{1}{483587934.400...};$$

(730) $$\frac{e^{5\Phi\pi e i_1 \sqrt{2}}}{483587934.400...} = \pi \frac{\Phi^3 i_1}{\text{Spin}} \cdot 10^{43.99975646300...}.$$

## Bioenergy
## &
## the Origin of Biological Evolution

Biological evolution up to the human genetic constitution becomes inevitable a process subject to the fundamental law

$$\widehat{E}\breve{E}\cdot\left\{\left[\!\left[\frac{\text{🕸}}{\text{😐}}\right]\!\right]\cdot\mathbf{Cr}_{+}^{\times}\cdot\mathbf{22727}\right\}=e^{5\Phi\pi i_1 e\sqrt{2}}, \qquad (731)$$

where the two, inert and animate, wings of crystallography are indispensible due to the ultra operator of geometry.

It is easy to see essentials of what is meant by bioenergy:

$$\frac{\{Genes\}_{+}^{\times}}{[\![\text{🕸}\text{😐}]\!]}\cdot\frac{\Delta_{Exprm.}}{\sin\Delta_{Exprm.}}=\mathbf{353}\cdot 10^{50}; \qquad (732)$$

$$\mathbf{353}\cdot\left\{\widehat{E}\breve{E}\cdot\mathbf{U}_{\mathbf{E}}\cdot E\right\}=\mathbf{7}\cdot 10^{39}.$$

$$\sqrt[Fluc]{\odot_{+}^{\times}\cdot\widehat{E}\breve{E}\cdot 65537}=\aleph_1\cdot 10^{41};$$

$$\sqrt[G\Pi]{\odot_{+}^{\times}\cdot e^{5\Phi\pi i_1 e\sqrt{2}}}=\mathbf{U}_{\mathbf{E}}\cdot 10^{18}.$$

$$\sqrt[\{Fluc\cdot Fluc\cdot Fluc\}]{\frac{\{Genes\}_{+}^{\times}}{[\![\text{🕸}\text{😐}]\!]}\left\{\widehat{E}\breve{E}\cdot\mathbf{U}_{\mathbf{E}}\cdot E\right\}}=\frac{\mathbf{2259496}}{9}\cdot 10^{12};$$

$$\mathbf{353}^{G\Pi\cdot G\Pi}=\exp\exp\tan 66\cdot 10^{29};$$

$$\mathbf{365.256}\cdot\pi\frac{\Phi^3\sqrt{i_1 i_2}}{\text{Spin}}\dim_{\Phi}\Phi^3 i_1(3+1)=100000;$$

$$\left\{\mathbf{365.256}\cdot\left\{\widehat{E}\breve{E}\cdot\mathbf{U}_{\mathbf{E}}\cdot E\right\}\right\}^{Fluc}=2\cdot 10^{68};$$

$$\sqrt[\{Fluc\cdot Fluc\cdot Fluc\}^3]{\odot_{+}^{\times}\frac{\{Genes\}_{+}^{\times}}{[\![\text{🕸}\text{😐}]\!]}\cdot\mathbf{365.256}}=\frac{10}{Fluc};$$

$$\left\{\mathbf{365.256}\cdot 12\right\}\left\{\widehat{E}\breve{E}\cdot\mathbf{U}_{\mathbf{E}}\cdot E\right\}\frac{\{Genes\}_{+}^{\times}}{[\![\text{🕸}\text{😐}]\!]}=\frac{10^{96}}{65537}. \qquad (741)$$

$$\left\{\odot_{+}^{\times}\cdot\{Genes\}_{+}^{\times}\cdot[\![\text{✵}\text{☺}]\!]\right\}\left\{\hat{E}\breve{E}\cdot\mathbf{U}_{\mathbf{E}}\cdot E\right\}=\frac{\mathbf{392783}}{3}\cdot 10^{177};$$

$$\sqrt[10]{\arccos\lg\lg\lg\lg\left\{\odot_{+}^{\times}\{\mathbf{365.256}\cdot 12\}\left\{\hat{E}\breve{E}\cdot\mathbf{U}_{\mathbf{E}}\cdot E\right\}\frac{3}{11}\right\}}=\Phi.$$

Consider universal energy with respect to the Solar system

(744) $$\sin\lg\left\{\sqrt[\Phi]{\mathbf{U}_{\mathbf{E}}}^{\pi e}\cdot\mathbf{42677}\right\}=\frac{\mathbf{1.618\,033\,988\,7...}}{2},$$

(745) $$\left\{\frac{\odot_{+}^{\times}\cdot\mathbf{42677}}{180\cdot 10^{47}}-1\right\}^{-\chi}=\frac{\mathbf{1982674}}{9}\cdot 10^{16};$$

(746) $$\mathbf{1982674}^{\Phi}=\frac{10^{11}}{\Phi\pi\sqrt{i_1 i_2}}.$$

The self-gravitating space-time dicontinuum is indispensible from universal energy

(747) $$\frac{G}{\Pi_{Cosm.}}\cdot\left\{\pi\frac{\Phi^{3}\sqrt{i_1 i_2}}{\text{Spin}}\cdot\mathbf{D}\{\mathbf{4}\}_{+}^{\times}\right\}\frac{1}{\mathbf{U}_{\mathbf{E}}}=\frac{\mathbf{2726}}{9}\cdot 10^{35.99999999...};$$

(748) $$\mathbf{2726}^{2\Phi\pi i_1}=\frac{\mathbf{241}}{9}\cdot 10^{43}, \textit{ et cetera.}$$

Electromagnetism is approachable as immediately as

(749) $$\sqrt[\Delta_{Exprm.}]{\sqrt[\Delta_{Exprm.}]{\mathbf{U}_{\mathbf{E}}}}=\Delta_1\cdot 10^{18}.$$

Just fluctuate the universal energy and derive the Coulomb charge made with Faraday's famous lines of forces which is the very source of Maxwell's equations (750):

$$\left\{\left\{\mathbf{U}_{\mathbf{E}}\cdot\left\{\left\{\frac{\dim E}{2}\right\}^{2}\right\}^{2}\right\}\cdot\left\{e^{\pm}\dim e^{\pm}\right\}^{2}\right\}^{\Phi\pi}\times\left\{\pi\frac{\Phi^{3}i_1}{\text{Spin}}\right\}^{-1}=10^{55}.$$

Electrodynamics as the main force acting in molecular biolody has many simple representations such as

(751) $$\left\{\left\{\mathbf{U}_{\mathbf{E}}\cdot\left\{\left\{\frac{\dim E}{2}\right\}^{2}\right\}^{2}\right\}\cdot\left\{e^{\pm}\dim e^{\pm}\right\}^{2}\right\}\cdot\Delta_{Exprm.}\equiv e^{\Phi}.$$

(752) $$\left\{\mathbf{U_E}\left\{e^{\pm}\dim e^{\pm}\right\}^2\frac{\Delta_{Exprm.}}{\sin\Delta_{Exprm.}}\right\}^{\chi}=\frac{\mathbf{86704213}}{9}\cdot 10^{57};$$

(753) $$\mathbf{86704213}\times e^{5\Phi\pi i_1}\cdot\frac{e^{5\Phi\pi e i_1\sqrt{2}}}{\widehat{E}\breve{E}}=\frac{10^{99.00000...}}{\mathbf{D\{4\}^{\times}_{+}}};$$

(754) $$\left\{\frac{1}{\varepsilon}\right\}^{\Phi\pi e}=\frac{\mathbf{3496171}}{3}\cdot 10^{60};$$

(755) $$\sqrt[e]{\mathbf{3496171}^{2\Phi\pi i_1}}=\mathbf{134985}\cdot 10^{26};$$

(756) $$\mathbf{134985}\cdot e^{\Phi\pi i_1}\cdot\frac{e^{5\Phi\pi e i_1\sqrt{2}}}{\widehat{E}\breve{E}}=\mathbf{32166}\cdot 10^{35},$$

*et cetera ad infinitum.*

One of many ways of grandunification will simply be

(757) $$\sqrt[\cos 2\Theta_{\textit{Strong Force}}]{\sin\Delta_{Exprm.}\sqrt{\sin\Theta_{\textit{Electroweak}}\sqrt{\mathbf{U_E}}}}=\frac{9}{\mathbf{670669}}\cdot 10^{83};$$

(758) $$\sqrt[\cos 2\Theta_{\textit{Strong Force}}]{\sin\Delta_{Exprm.}\sqrt{\cos\Theta_W\sqrt{\sin\Theta_{\textit{Electroweak}}\sqrt{\mathbf{U_E}}}}}=\frac{10^{87}}{\sin\Theta_W}.$$

Since crystallographic symmetries provide a localizing operator such that

(759) $$\left\{\mathbf{Cr^{\times}_{+}}\right\}^{\pi}=\mathbf{22727}\cdot 10^{55},$$

human genetics in terms of universal energy reduces to

(760) $$\left\{[\![\text{🕸☺}]\!]\mathbf{Cr^{\times}_{+}}\cdot\mathbf{22727}\right\}\cdot\mathbf{U_E}=\frac{10^{62}}{\mathbf{26329}}.$$

This implies that the animated universal energy acquires its own operator, 26329, for the purpose of whatever bioenergy functions possible in the 4-space-time medium. It will, first and foremost, emanate something like Faraday's lines of forces:

(761) $$\mathbf{26329}\cdot e^{\Phi\pi i_1}=\pi\Phi^3 i_1\cdot 10^6;$$

(762) $$\mathbf{26329}\cdot e^{\Phi\pi i_1}\cdot\frac{e^{5\Phi\pi e i_1\sqrt{2}}}{\widehat{E}\breve{E}}=\mathbf{627402}\cdot 10^{33},$$

giving birth to a new universal operator.

Boundaries of the above operator's influence are defined by the surface area of a parallelopiped with measurements

(763) $$\{\Phi,\ 1,\ \Phi^{-1}\},$$

as it can be seen in

(764) $$\sqrt[\dim_\Phi \Phi^3 i_1]{\sqrt{\mathbf{627402}^{\pi^{\frac{\Phi^3 i_1}{\text{Spin}}}}}} = \frac{10^{17}}{4\Phi}\left\{1+\frac{1}{\mathbf{Z}}\right\}.$$

Circumscribe it by a sphere and get another surface area

(765) $$4\Phi \to 4\pi,$$

as well as its consequence such that

(766) $$\sqrt[G\Pi]{\mathbf{627402}^{4\pi}} = \frac{10^{25}}{\mathbf{61482}}.$$

Now that we have to go va-banque in this ultimate game and see what is the profit

(767): $$\mathbf{627402}\cdot\{Genes\}_{+}^{\times} = \mathbf{1040082}\cdot 10^{75}.$$

(768) $$\mathbf{1040082}\cdot e^{\Phi\pi i_1} \equiv 2\Theta_{\textit{Strong Nuclear Force}}.$$

Moreover,

(769) $$\mathbf{1040082}\cdot\left\{\odot_{+}^{\times}\cdot 🕸\right\}\times\mathbf{9} = \frac{10^{68.99999...}}{\mathbf{5.71100522647...}},$$

implying that simply by definition:

(770) $$\mathbf{1040082}\cdot\left\{\odot_{+}^{\times}\cdot 🕸\right\}\times\mathbf{9} \equiv \exists.$$

We have general configurations such as

(771) $$\mathbf{1040082}\cdot\left\{\odot_{+}^{\times}\cdot[\![ 🕸 ⊜ ]\!]\left\{\mathbf{Cr}_{+}^{\times}\cdot\mathbf{22727}\right\}\right\} = \frac{10^{103}}{\mathbf{13565}};$$

(772) $$\frac{🕸}{\text{Spin}} = (©@)^2\cdot 10^{18}.$$

The next fundamentality cannot be surpassed

(773) $$\frac{🕸\cdot\sqrt[\Phi\pi e+1]{10^{90}}}{\text{Spin}} = \mathbf{607025}\cdot 10^{17};$$

(774) $$\frac{\mathbf{607025}\times\mathbf{U_E}}{\Delta_1\times\left\{\mathbf{360}\cdot\{\mathbf{3\cdot 5\cdot 17\cdot 257\cdot 65537}\}\cdot\Phi\pi i_1\right\}} = 1-\frac{1}{1000\mathbf{U}}.$$

## The Real-Imaginary Ultra Operator

It is necessary to compose

(775) $$\exp(\sin i \cos i \times 5\Phi \pi i_1 e\sqrt{2})$$

and see how does the real-imaginary ultrafine operator unify physics:

(776) $$e^{\cos i \cdot \sin i \times 5\Phi \pi e i_1 \sqrt{2}} \cdot \left\{ \left\{ \frac{Ghm_e e^{\pm} c \cdot \dim(Ghm_e e^{\pm} c)}{\alpha a_e} \right\} \Theta_W \Theta_{STR.} \right\} = \left\{ \frac{\mathbf{48265}}{9} \right\}^2 \cdot 10^{103.00000000...}.$$

Here the real-imaginary ultrafine operator is

$$\mathbf{7.5937031282007092382806120639486 \cdot 10^{97}},$$

and the experimental constants and the new parameters arrive at

$$\mathbf{3787267587620.7375136109523035562}.$$

After all this mess the only fuss will be just to know that there is no more fundamentality in pure maths than that of Gauss's theory and that the superunified cosmic field is such as it is next:

(777) $$\frac{\left\{ \left\{ \frac{Ghm_e e^{\pm} c \cdot \dim(Ghm_e e^{\pm} c)}{\alpha a_e} \right\} \Theta_W \Theta_{STR.} \right\}}{\sin \Delta_{Exprm.} \sin \Theta_W \cos \Theta_W \cos 2\Theta_{STR.}} = \frac{10^{28}}{\mathbf{\{3,5,17,257,65537\}_{+}^{\times}}}.$$

We do shift to entirely new paradigms; thus, never underestimate resources of unification solutions in pure mathematics such as

(778) $$\sqrt[e]{\left\{ \sqrt[G\Pi]{\left\{ \sqrt[\otimes]{\{\mathbf{48265}\}^{\oplus}} \right\}^{\chi}} \right\}^{\pi}} = \frac{10^{15}}{\sqrt{2}}.$$

Indeed, maths began with the Pythagoras' constant, though this man drowned the author of the discovery saying that irrationalities do not exist in mathematics. What does only exist is the mathematical continuum self-organized into the system of universal harmony where do count, indeed, integers alone.

## The Moon Cycles and Life

The Sun shines in the day light, while the Moon makes dark nights glow. Needless to say, therefore, that Moon's astronomy is much brighter mathematically than that of the Sun. In this respect are exemplary the Saros and Metonic cycles, in particular.

In the unifield theory we do always refer to the webcirculated astronomical data by James Q. Jacobs:

$$242 \times 27.21222 = 6585.357425$$
$$223 \times 29.53059 = 6585.321321$$
$$239 \times 27.55455 = 6585.536614$$
$$19 \times 346.62006 = 6585.781197$$

We find first that Saros cycle is as fundamental as next

(779) $$\{\mathbf{Saros}\}\cdot e^{5\Phi\pi i_1 e\sqrt{2}} = \frac{\Phi^3 i_1}{\text{Spin}}\cdot 10^{57}.$$

Therefore, theoretically

(780) $$\{\mathbf{Saros}\} = 6585.94664.$$

The mean observational duration of the Saros cycle owes to the self-perturbation effect of geometry defined by

(781) $$\frac{1}{\varepsilon}\cdot G = \frac{\mathbf{8838092}}{9\cdot 10};$$

(782) $$\mathbf{8838092}\times\{\Phi\pi i_1 e\}\cdot\{\oplus\otimes\}\cdot\{\chi\cdot G\Pi\}\cdot e^{5\Phi\pi i_1 e\sqrt{2}}\cdot e^{10\Phi\pi i_1} = \Delta_1\cdot 10^{92}.$$

This prompts us to guess immediately that

(783) $$\{\{\mathbf{Saros}\} = 6585.94664\}\cdot\circledast \equiv 100000\Psi,$$

where we refer to the totally unified configuration of physics.

(784) $$\Psi = \frac{\left\{ \begin{array}{c} \frac{Ghm_e e^{\pm} c \cdot \dim(Ghm_e e^{\pm} c)}{\alpha a_e} \cdot \Theta_W \Theta_{STR.} \cdot \\ \cdot \{\{2 \times 2 \cdot 6\} \cdot \boldsymbol{FB}\} \end{array} \right\}}{\sin \Delta_{Exprm.} \sin \Theta_W \cos \Theta_W \cos 2\Theta_{STR.}} =$$

$$= \mathbf{2.920925121} \cdot 10^{15}.$$

Consequently, we are entitled to explain the reason for the life existence by the following system of equations:

(785) $$\Psi \cdot 😐 = \frac{\Phi^3 i_1}{\text{Spin}} \cdot 10^{22.999...};$$

(786) $$\Psi [\![ 🕸 😐 ]\!] = \frac{10^{46.000...}}{\&^{\times}};$$

(787) $$\Psi \cdot \odot_{+}^{\times} = \sqrt[\Phi]{\pi \Phi^3 i_1 \cdot 10^{96.0000...}};$$

(788) $$\sqrt[\Phi]{\Psi \cdot \odot_{+}^{\times}} = \Delta_1 \cdot 10^{34.999...};$$

(789) $$\{\mathbf{Saros}\} \cdot \odot_{+}^{\times} = \frac{10^{51.0000...}}{360}.$$

On the other hand,

(790) $$\{242, 223, 239, 19\}_{+}^{\times} \cdot 🕸 = \frac{10^{27.9999...}}{\sqrt{i_1 i_2}}.$$

The Metonic cycle reads as below

235 x 29,53059 = 6939,688388

255 x 27,21222 = 6939,116295

19 x 365,24219 = 6939,601660

where we will certainly have the same picture as above (16)

(791) $$\frac{\Psi}{\{\mathbf{Metonic}\}} = \frac{9}{21382} \cdot 10^{15},$$

whence theoretically

(792) $$\left\{\mathbf{Metonic}\right\} = 6939.468993$$

with the consequence that

(793) $$\left\{\mathbf{Metonic}\right\}\text{⚝} = \tan 72 \cdot 10^{19.9999...}.$$

We could somewhat predict that

(794) $$\{\mathbf{Saros, Metonic}\}_{+}^{\times} = \frac{10^{12.0000...}}{\Phi};$$

(795) $$\{\mathbf{Saros, Metonic}\}_{+}^{\times} \cdot e^{5\Phi\pi i_1 e\sqrt{2}} \times \left\{\left\{\frac{\dim E}{2}\right\}^{2}\right\}^{2} = 10^{66.000...}.$$

Note that

(796) $$\frac{^{\mathbf{THEOR.}}\{\mathbf{Saros, Metonic}\}_{+}^{\times}}{^{\mathbf{OBSERV.}}\{\mathbf{Saros, Metonic}\}_{+}^{\times}} = 1 + \sqrt[\chi\cdot G\Pi]{\frac{\alpha a_e}{10^{70.9999...}}};$$

(797) $$^{\mathbf{OBSERV.}}\{\mathbf{Saros, Metonic}\}_{+}^{\times} = \frac{\mathbf{6260881600}}{\mathbf{137}}.$$

Cosmologically,

(798) $$^{\mathbf{OBSERV.}}\{\mathbf{Saros, Metonic}\}_{+}^{\times} \cdot \odot_{+}^{\times} \cdot \mathbf{U}_{\mathbf{E}} = \frac{10^{66.00000...}}{\pi\Pi}.$$

Since we know that

(799) $$[\![\text{⚝}\text{☺}]\!] \cdot \left\{\mathbf{Cr}_{+}^{\times} \cdot \mathbf{22727}\right\} = \mathbf{168038} \cdot 10^{180-137},$$

we have to have now

(800) $$^{\mathbf{OBSERV.}}\{\mathbf{Saros \times Metonic}\} \cdot \mathbf{168038} = \Phi^{\Phi\Phi\Phi} \cdot 10^{12.0000...};$$

(801) $$\frac{^{\mathbf{OBSERV.}}\{\mathbf{Saros \times Metonic}\}}{\alpha a_e} = \Pi \cdot 10^{12.999...}.$$

In the end everything under the Sun comes from the Gaussian extension of the Euclidean system:

(802) $$\left\{^{\mathbf{OBSERV.}}\{\mathbf{Saros \times Metonic}\} \cdot \mathbf{U}_{\mathbf{E}}\right\}^{\pi} = \\ = \{\mathbf{3, 5, 17, 257, 65537}\}_{+}^{\times} \cdot 10^{39}.$$

The lesson here is that the Moon cycles are based upon a constructive algorithms and, therefore, they do serve a most energizing factor to be studied further.

So, in the four-dimensional topology of space-time the Cosmological Trinity of life existence is composed of the constructive algorithms, universal energy and gene code:

(803) $$\{\mathbf{3,5,17,257,65537}\}^{\times}_{+} \cdot \mathbf{U_E} \cdot \text{🕸} = \frac{\mathbf{D\{4\}^{\times}_{+}}}{100000}.$$

It is true that the Moon does exist to wake up and inspire the human intelligence. Yet, the latter is a proverbially ackward and awfully inept phenomenon: I always knew that something like to the following should there be, but it took incredibly long years to derive this brilliance just now:

(804) $$\{\mathbf{3,5,17,257,65537}\}^{\times}_{+} \cdot \mathbf{U_E} \cdot \odot^{\times}_{+} = \frac{\mathbf{254614}}{e^{5\Phi\pi i_1 e\sqrt{2}}} \cdot 10^{117};$$

(805) $$\mathbf{254614}^{\pi\Phi^3 i_1} = \frac{\mathbf{97297}}{3} \cdot 10^{87};$$

(806) $$\mathbf{97297}^{\Phi\pi i_1 e} = \frac{\mathbf{42290561}}{9} \cdot 10^{81}.$$

As the reader directly witnesses, all the above findings owe to the Moon's inspiring influence. The Solar system of planets is an ultimate mathematical machine working with an accuracy up to

$$e = 2.718\ 281\ 828\ 460 \cdot$$

Saros' self-reference is

(807) $$\{\{\mathbf{Saros}\} = 6585.94664\} \cdot$$
$$\cdot \{254614 \cdot 97297 \cdot 42290561\} = \mathbf{68.999...} \cdot 10^{20},$$

because of the dynamical equilibrium of symmetries of geometry

(808) $$\frac{1}{2}\left\{\frac{360}{\cosh\cos^2 i} + \frac{360}{5}\right\} = 68.999...$$

So, the Moon writes the following life-and man equations:

(809) $$\{\{\mathbf{Saros}\} = 6585.94664\} \cdot \{254614 \cdot 97297 \cdot 42290561\} \cdot$$
$$\cdot \{[\![\text{🕸}]\!] \cdot \mathbf{Cr}^{\times}_{+} \cdot \mathbf{22727}\} = \frac{10^{62}}{\cos(180 - \Phi^{10})};$$

(810) $$\{\{\mathbf{Saros}\} = 6585.94664\} \cdot \{254614 \cdot 97297 \cdot 42290561\} \cdot$$
$$\cdot \{[\![\text{🕸😐}]\!] \cdot \mathbf{Cr}^{\times}_{+} \cdot \mathbf{22727}\} = \frac{10^{73}}{\aleph_1}.$$

## Duerer's & Ramanujan's Magic Squares as Cosmogonic Codes

The real philosophy is never spoken much. It is simply sort of idiotism if an idle man writes many volumes of pure speculative philosophy where you never know what the man wants to say. The best of German philosophy and, possibly, of all philosophies had been given by Albrecht Duerer in his most famous copper engraving Melancholia I as early as 500 years ago.

It is one of most mystical works in the art history and as such it collected a Mont-Blanc of literature. But, the true interpretation comes if only time comes. Secondly, such works of due greatness are usually inspired by hidden sources. In the given case Duerer was not obliged to know what purpose he had been employed for.

The Melancholia renders man's desparate desire for the complete knowledge despite the weakness of human intelligence.

Duerer's magic square is

(811)

| | | | |
|---|---|---|---|
| 16 | 3 | 2 | 13 |
| 5 | 10 | 11 | 8 |
| 9 | 6 | 7 | 12 |
| 4 | 15 | 14 | 1 |

$$\left.\right._{+}^{\times} = \{\mathbf{D}_{MS}\}_{+}^{\times}$$

and immediately find that

(812) $$\{\mathbf{D}_{MS}\}_{+}^{\times\exists} = \sqrt[\text{©@}]{\mathbf{93917}\cdot 10^{15}}\,,$$

which a very precise expression.

The Ramanujan magic square is

(813)

| | | | |
|---|---|---|---|
| 22 | 12 | 18 | 87 |
| 88 | 17 | 9 | 25 |
| 10 | 24 | 89 | 16 |
| 19 | 86 | 23 | 11 |

$$\left.\right._{+}^{\times} = \{\mathbf{R}_{MS}\}_{+}^{\times}$$

and

(814) $$\sqrt[\Phi]{\{\mathbf{R}_{MS}\}_{+}^{\times\pi}} = \mathbf{4124529}\cdot 10^{42}\,.$$

If explicitly,

(815) $$\sqrt[\Phi]{\{1093146559376971795660800\}^{3.141\,592\,653\,59}} = \\ = \mathbf{4124529} \cdot 10^{42}.$$

And fundamentally

(816) $$\{\mathbf{R_{MS}}\}_{+}^{\times} \cdot \{\mathbf{3,5,17,257,65537}\}_{+}^{\times} = \cos\frac{2\pi}{5} \cdot 10^{40}.$$

This means that we have to combine the two glorious names of Gauss and Ramanujan in the following unified configuration

(817) $$\left\{\sqrt{\frac{\pi e}{2}}\right\}^{\Phi} \cdot \{\mathbf{R_{MS}}\}_{+}^{\times} \cdot \{\mathbf{3,5,17,257,65537}\}_{+}^{\times} \equiv \mathbf{1}.$$

As one could snear, the problem becomes even more magical and promising. Thus, remind that (818)

$$\text{Physics} = \frac{\left\{\begin{array}{c} \dfrac{Ghm_e e^{\pm} c \cdot \dim(Ghm_e e^{\pm} c)}{\alpha a_e} \cdot \Theta_W \Theta_{STR.} \cdot \\ \cdot \{\{2 \times 2 \cdot 6\} \cdot \boldsymbol{FB}\} \end{array}\right\}}{\sin \Delta_{Exprm.} \sin \Theta_W \cos \Theta_W \cos 2\Theta_{STR.}} = \\ = \mathbf{2.920925121} \cdot 10^{15}.$$

and be brave enough to go va-banque and see a consistent

(819) $$\{\mathbf{R_{MS}}\}_{+}^{\times} \cdot \text{Physics} = \mathbf{3193} \cdot 10^{37};$$

(820) $$\mathbf{3193}^{\chi} = \frac{10^{20}}{\mathbf{180}}.$$

The rest we leave to the readers' curiosity and exersizes.

## The Imperial Code

As the Marquise Pierre-Simon Laplace highlighted in his famous treatise **Exposition du Systeme du Monde**, during the Mongolian Empire astronomy and mathematics were revived and flowerished in every corner of the then known half of the Globe. A magic square was composed and cast in iron on the occasion of the foundation of the first Academy of sciences by the Great Khaan Khubilai. In the Western literature is most famous and mysterious Coleridge's poem devouted to this Emperor. Therefore, we may name it the Imperial Mongol magic square and write its topology as usually

(821)

| | | | | | |
|---|---|---|---|---|---|
| 28 | 4 | 3 | 31 | 35 | 10 |
| 36 | 18 | 21 | 24 | 11 | 1 |
| 7 | 23 | 12 | 17 | 22 | 30 |
| 8 | 13 | 26 | 19 | 16 | 29 |
| 5 | 20 | 15 | 14 | 25 | 32 |
| 27 | 33 | 34 | 6 | 2 | 9 |

$$= \{\mathbf{IM}_{\mathbf{MS}}\}^{\times}_{+}.$$

It is predictable that

(822) $$\{\mathbf{IM}_{\mathbf{MS}}\}^{\times}_{+}\cdot\textbf{Physics}=\frac{10^{61}}{\Phi\pi e};$$

(823) $$\frac{\left\{\begin{array}{c}\dfrac{Ghm_e e^{\pm}c\cdot\dim(Ghm_e e^{\pm}c)}{\alpha a_e}\cdot\Theta_W\Theta_{STR.}\cdot\\ \cdot\{\{2\times2\cdot6\}\cdot\boldsymbol{FB}\}\end{array}\right\}}{\sin\Delta_{Exprm.}\sin\Theta_W\cos\Theta_W\cos2\Theta_{STR.}}=\textbf{Physics}.$$

If take unified physics in complete, then

(824) $$\sqrt[\pi]{\left\{\{\mathbf{IM}_{\mathbf{MS}}\}^{\times}_{+}\cdot\left\{\text{Physics}\cdot\frac{\exists}{\copyright @}\cdot\mathbf{92}\right\}\right\}^{e}}=\Phi\pi\cdot10^{54}.$$

Derive the Maxwell's constant from

(825) $$\{\mathbf{IM}_{\mathbf{MS}}\}_{+}^{\times}\cdot\mathbf{299792458}i_1=\frac{e^{5\Phi\pi e i_1\sqrt{2}}}{9.999...}.$$

Then, what is life?

(826) $$\{\mathbf{IM}_{\mathbf{MS}}\}_{+}^{\times}\cdot\left\{\text{🕸}\cdot\mathbf{Cr}_{+}^{\times}\cdot\mathbf{22727}\right\}=\pi\frac{\Phi^{3}\sqrt{i_1 i_2}}{\text{Spin}}\cdot 10^{83};$$

(827) $$\{\mathbf{IM}_{\mathbf{MS}}\}_{+}^{\times}\cdot\left\{\odot_{+}^{\times}\cdot\text{😐}\right\}=\nabla\cdot 10^{95}.$$

What do alone exist is numbers and logoses such as

(828) $$\frac{\{\mathbf{IM}_{\mathbf{MS}}\}_{+}^{\times}}{\{\mathbf{D}_{\mathbf{MS}}\}_{+}^{\times}\cdot\{\mathbf{R}_{\mathbf{MS}}\}_{+}^{\times}}=\sqrt[\pi]{\frac{\mathbf{1622323}}{9}\cdot 10^{7}};$$

(829) $$\sqrt[\Phi\pi i_1]{\left[\!\left[\{\mathbf{IM}_{\mathbf{MS}}\}_{+}^{\times}\cdot\{\mathbf{D}_{\mathbf{MS}}\}_{+}^{\times}\cdot\{\mathbf{R}_{\mathbf{MS}}\}_{+}^{\times}\right]\!\right]^{e}}=\mathbf{48581}\cdot 10^{31}.$$

So, anyway, the greatest trinity is by Providence (830)

| | | | | | |
|---|---|---|---|---|---|
| **28** | **4** | **3** | **31** | **35** | **10** |
| **36** | **18** | **21** | **24** | **11** | **1** |
| **7** | **23** | **12** | **17** | **22** | **30** |
| **8** | **13** | **26** | **19** | **16** | **29** |
| **5** | **20** | **15** | **14** | **25** | **32** |
| **27** | **33** | **34** | **6** | **2** | **9** |

| | | | |
|---|---|---|---|
| 16 | 3 | 2 | 13 |
| 5 | 10 | 11 | 8 |
| 9 | 6 | 7 | 12 |
| 4 | 15 | 14 | 1 |

| | | | |
|---|---|---|---|
| 22 | 12 | 18 | 87 |
| 88 | 17 | 9 | 25 |
| 10 | 24 | 89 | 16 |
| 19 | 86 | 23 | 11 |

Therefore, we have to have

(831) $$\mathbf{48581}\left\langle\odot_{+}^{\times}\left[\!\left[\text{🕸}\text{😐}\right]\!\right]\right|=\mathbf{1934919}\cdot 10^{68}.$$

Finally,

(832) $$\left\{\begin{array}{l}\dfrac{2N\cdot Ghm_e e^{\pm}c\cdot\dim(Ghm_e e^{\pm}c)}{\alpha a_e}\cdot\\ \cdot\dfrac{\Theta_W\Theta_{STR.}\{2\times 2\cdot\mathbf{6}\cdot\boldsymbol{FB}\}\times\mathbf{92}}{\sin\Delta_{Exprm.}\sin\Theta_W\cos\Theta_W\cos 2\Theta_{STR.}}\end{array}\right\}\mathbf{48581}=$$

$$=\sqrt[\Phi]{\frac{\mathbf{10711975}}{3}\cdot 10^{35}}.$$

## Quantum Gyroscope
## Pro & Contra Supercollider

> Beauty is a wider, and more fundamental, notion than Truth... The teleology of the Universe is directed to the production of Beauty... The Truth of supreme Beauty lies beyond the dictionary meanings of words...Truth, Beauty, and Goodness.
>
> **Alfred North Whitehead**

It is simply that

(833) $$\Delta_{Exprm.}\frac{G}{\Pi} \equiv \pi\Phi^3 i_1 .$$

Accurately,

(834) $$\left\{\Delta_{Exprm.}\frac{G}{\Pi}\right\} = 100\pi\Phi^3 i_1\left\{1+\frac{1}{45888.00676}\right\};$$

(835) $$45888.00676\times\mathbf{65537} = \mathbf{3007362300}\cdot$$

If otherwise,

(836) $$\left\{\Delta_{Exprm.}\frac{G}{\Pi}\right\}^{\oplus} = \mathbf{2745}\cdot 10^{26};$$

(837) $$\mathbf{2745}\cdot\pi\frac{\Phi^3\sqrt{i_1 i_2}}{\text{Spin}}\cdot\dim_{\Phi}\Phi^3 i_1\cdot(3+1)\cdot\mathbf{D\{4\}}_{+}^{\times} = \frac{\mathbf{1916783}}{9}\cdot 10^{46}.$$

Now interpret the following system:

(838) $$Nm_e \dim m\cdot e^{\pm}\dim e^{\pm} = \frac{10^7}{\Delta_{Exprm.}};$$

(839) $$\sqrt[e]{\left\{\frac{10^7}{\Delta_{Exprm.}}\cdot\frac{G}{\Pi}\right\}^{\Phi\pi i_1}} = \frac{\mathbf{1315777}}{9}\cdot 10^9;$$

(840) $$\sqrt[e]{\left\{\frac{10^7}{\Delta_{Exprm.}}\cdot\frac{G}{\Pi}\right\}^{\Phi\pi i_1}}\cdot 9\cdot\mathbf{65537}=\frac{10^{17}}{a_e}.$$

**What is the cause for what here?** Here and elsewhere in the absolute geometry and, accordingly, in Nature everything existent is the cause for all the rest of phenomena and *vice versa*: all the phenomena in their totality do determine every single phenomenon given. And in this system overrules the fundamental algebraic bifurcation

(841) $$\cos i \rightleftarrows \cos 2\Theta_{\textit{Nuclear Strong Force}}.$$

Applying to the installed *Microsoft* calculator we will have

(842) $$1-arch\frac{\left\{10\cos 2\Theta_{\textit{Nuclear Strong Force}}\right\}^2}{10}=$$
$$=\frac{1}{\mathbb{Z}=456746.3616321157459582867002782 6};$$

(843) $$\mathbb{Z}^{2\Phi\pi e i_1}=\cos\Theta_W\cdot 10^{\mathbf{198.99999}77207828741869346842491 4}.$$

The system logic of the universal harmony is based upon the pentasymmetric mechanism such that:

**Duality of Beauty and Truth;**

**Trinity of Heuristics, Aphoristics and Mnemonics.**

Such is the genuine formalism and the liberal art of both the divine and human intelligence. If so, there should exist an equation that equals to

(844) $$\pi\frac{\Phi^3 i_1}{\text{Spin}}\cdot 10^{10}.$$

Knowing some subtlest mechanics of geometry, we find

(845) $$\sqrt[\sin\Delta_{Exprm.}]{\mathbb{Q}\cdot\frac{\Delta_{Exprm.}}{\sin\Delta_{Exprm.}}}=\pi\frac{\Phi^3 i_1}{\text{Spin}}\cdot 10^{10};$$

(846) $$\mathbb{Q}=\mathbf{246581.9476...}$$

If believe the very logic of our geometry, this Q has to have a meaning. Then,

(847) $$\sqrt[\cos\Theta_W]{\left\{\sqrt[\sin\Theta_W]{\mathbb{Q}\cdot\frac{\Theta_W}{\sin\Theta_W}}\right\}\frac{1}{\cos\Theta_W}}\times 2\Theta_{STR.}=\mathbf{622}\cdot 10^{17};$$

(848) $$\mathbb{Q}=\mathbf{246580.3665...}$$

Now that **a self-referencial equation** written by the same parameter of the nuclear strong interaction

(849) $$\sqrt[\cos 2\Theta_{STR.}]{2\Theta_{STR.}\left\{\frac{\mathbb{Q}}{\cos 2\Theta_{STR.}}\right\}}=\frac{10^{19}}{\cos 2\Theta_{STR.}};$$

(850) $$\mathbb{Q}=\mathbf{246580.2518...}.$$

The key word here is 'self-reference' and we have accurately

(851) $$\mathbb{Q}=246580.2518227964895992989368 8149.$$

It is an operator designed to provide something like

(852) $$\mathbb{Q}\cdot m_e(0.511MeV)=126002.50868144900...GeV?$$

If try in terms of fundamental mass spectra, then

(853) $$\frac{\{\mathbb{Q},N_{1837.41}\}_{+}^{\times}}{m_e(0.511MeV)}=e^{10}\cdot 10^{10};$$

(854) $$\{\mathbb{Q},N\}_{+}^{\times}\times\mathbf{U}_{\mathrm{E}}=\frac{10^{23}}{\cos 2\Theta_{STR.}}.$$

So, fundamentally

(855) $$\mathbb{Q}\rightarrow Boson;\quad N\rightarrow Fermion?$$

If 'Yes', then

(856) $$\mathbb{Q}\cdot m_e(0.511MeV)=m_{\mathbf{Higgs\ boson}}?$$

Why not, for

(857) $$\{\mathbb{Q}\cdot N\}\cdot m_e\dim m\times\left\{\left\{\frac{\dim E}{2}\right\}^2\right\}^2=10^{10};$$

(858) $$\left\{\sqrt{2}\cdot\{\mathbb{Q},N\}_{+}^{\times}\right\}=\frac{10^{15}}{2\pi};$$

(859) $$\left\{e\{\mathbb{Q},N\}_{+}^{\times}\right\}^{e}=\frac{\mathbf{213943}}{9}\cdot 10^{35};$$

(860) $$\left\{\chi\{\mathbb{Q},N\}_{+}^{\times}=57\cdot 10^{13}\right\}^{\chi}=\frac{10^{76}}{19},$$

*et cetera*, and so on.

The following sheds much light on the whole problem

(861) $$e^{5\Phi\pi e i_1\sqrt{2}}\cdot\{\mathbb{Q}\cdot N\}\cdot \boldsymbol{F}^2=\tan 66\tan 72\cdot 10^{62};$$

(862) $$e^{5\Phi\pi e i_1\sqrt{2}}\cdot\{\mathbb{Q}\cdot N\}\cdot \boldsymbol{B}=a_e\cdot 10^{66};$$

(863) $$\tan 66\tan 72\cdot a_e=\frac{\mathbf{80162}}{10000000},$$

because of yet another fundamental bifurcation such that

(864) $$\left\{\frac{360}{\cosh\cos^2 i}\cong 66\right\}\Leftrightarrow\left\{\frac{2\pi}{5}\cong 72\right\}.$$

The Unifield theory is an art to approximate as many constants as possible in one go and thus a very precise

(865) $$e^{\sin i\cos i\times 5\Phi\pi e i_1\sqrt{2}}\cdot\tan\mathbf{66}\tan\mathbf{72}\cdot\mathbf{U_E}=\frac{9}{\mathbf{75856}}\cdot 10^{112}.$$

Therefore, we apply to the installed Microsoft calculator and obtain the universal energy configuration as

(866) $$\mathbf{U_E}^*=2260258612.6408351872909333035202,$$

while the same configuration with the experimental values equals to

(867) $$\mathbf{U_E}=2260258601.268245380...=$$
$$=E_{ag}\cdot hh_{ag}\cdot 2\mathrm{N}m_e c^2\cdot\frac{2\Theta_{STR.}}{\cos 2\Theta_{STR.}}.$$

Then, the question arises: Whether the perturbation effect

(868) $$\frac{1}{\mathbb{C}=198746164.2043222...}$$

that breaks here symmetries has any sense or not ?

It makes sense (869)

$$\frac{1}{360}\lg\frac{\mathbb{C}\cdot\pi^{\frac{\Phi^3\sqrt{i_1 i_2}}{\mathrm{Spin}}}\cdot\dim_\Phi\Phi^3 i_1\cdot(3+1)\cdot\mathbf{D\{4\}}_+^\times}{\cos i}=\frac{100}{e^{\Phi\pi i_1}}.$$

Approximate the following not quite just for fun:

(870) $$\sqrt{\left\{\sqrt[i_1]{\left\{\sqrt[e]{\left\{\sqrt[\mathrm{GTI}]{\left\{\sqrt[\otimes]{\{\mathbf{75856}\}^{\oplus}}\right\}^{\chi}}\right\}^{\pi}}\right\}^{\Phi}}\right\}^{5}}=\frac{10^{50}}{2\pi}.$$

Whatever the case, Q is a cosmogically fundamental parameter because of the standard Φ-test of harmony (871):

$$\sqrt[10]{\arccos \lg\lg\lg\lg \frac{\mathbb{Q}\cdot\pi\frac{\Phi^3\sqrt{i_1 i_2}}{\text{Spin}}\cdot \dim_\Phi \Phi^3 i_1\cdot(3+1)\cdot \mathbf{D}\{2\}_+^\times\cdot\mathbf{D}\{3\}_+^\times\cdot\mathbf{D}\{4\}_+^\times}{\pi\Pi}}=\Phi\cdot$$

So, a general energy-mass related theorem will read physically

(872) $$\sqrt[\pi]{\exp\exp e\cdot 10^{50}}=\mathbb{Q}\cdot N\cdot\mathbf{U}_\mathbf{E},$$

and mathematically

(873) $$\frac{\mathbb{Q}}{N}=\sqrt[\chi\chi]{\mathbf{3639}\cdot 10^{51}},$$

and in both terms

(874) $$\mathbf{3639}\cdot\cdot\pi\frac{\Phi^3\sqrt{i_1 i_2}}{\text{Spin}}\cdot\dim_\Phi \Phi^3 i_1\cdot(3+1)\cdot \cdot\frac{\mathbf{D}\{2\}_+^\times\cdot\mathbf{D}\{3\}_+^\times\cdot\mathbf{D}\{4\}_+^\times}{\mathbf{InG}_+^\times}\cdot\mathbf{U}_\mathbf{E}=\mathbf{206414}\cdot 10^{74}.$$

And Cosmo-Logically

(875) $$\{\mathbb{Q}\cdot N\}\cdot \boldsymbol{FB}=\sqrt[G\Pi]{\Delta_\alpha\cdot 10^{31}}\,.$$

We have the masses of the three intermediary bosons such that

(876) $$\frac{M_{Z(0)}+M_{W(+)}+M_{W(-)}}{m_e}\cdot\sqrt{2}=\mathbf{697375}\,;$$

(877) $$\mathbf{697375}\cdot\pi\frac{\Phi^3\sqrt{i_1 i_2}}{\text{Spin}}=\frac{\mathbf{40913983}}{3}.$$

Now we may complete the configuration of the boson mass spectrum with what can be Higgs boson and have (878)

$$\left\{\frac{126002.5087MeV+\{2\times 80398MeV+91187.6MeV\}}{0,511MeV}\right\}=\mathbf{M}_B\,;$$

(879) $$\mathbf{M}_B=739698.8429\,.$$

So, if primitively

(880) $$\mathbf{M}_B{}^{\Phi\Phi\Phi} = \frac{10^{27}}{\Delta_1},$$

flattering those who believe that all masses owe to electromagnetism.

Or even

(881) $$\sqrt[\sin\Delta_{Exprm.}]{\sqrt[\sin\Delta_{Exprm.}]{\mathbf{M}_B \cdot \frac{\Delta_{Exprm.}}{\sin\Delta_{Exprm.}}}} = \cos 2\Theta_{STR.} \cdot 10^{18}.$$

Now that the following exact formula is well predictable:

(882) $$\mathbf{M}_B \cdot \frac{2\Theta_{STR.}}{\cos 2\Theta_{STR.}} = \frac{\mathbf{1133272000}}{9}.$$

It is foreseeable that also

(883) $$\mathbf{M}_B \cdot \frac{e^{5\Phi\pi e i_1\sqrt{2}}}{\overset{\frown}{E}\breve{E}} = \mathbf{InG}_{+}^{\times} \cdot 10^{24}.$$

The following is a logical expectation, too,

(884) $$\mathbf{M}_B \cdot \boldsymbol{FB} = \frac{\mathbf{22910000}}{9}.$$

The configuration is, indeed, something absolute and ultimate, for

(885) $$\mathbf{M}_B = \frac{\sqrt[\Phi]{\mathbf{6830924400}}}{\Phi}.$$

The mass formation mechanism of boson matter can be seen from such monsterous but consistent expressions like

(886) $$\left\{M_{Z(0)} \cdot m_e\right\} \cdot \frac{e^{5\Phi\pi e i_1\sqrt{2}}}{\overset{\frown}{E}\breve{E} \cdot \mathbf{U}_{\mathbf{E}} \cdot E} = \mathbf{222} \cdot 10^{19};$$

(887) $$\sqrt{\left\{M_{W(\pm)} \cdot m_e\right\} \cdot \frac{e^{5\Phi\pi e i_1\sqrt{2}}}{\overset{\frown}{E}\breve{E} \cdot \mathbf{U}_{\mathbf{E}} \cdot E}} = \frac{10^{20}}{\mathbf{U}_{\mathbf{E}}}.$$

At least in this essay I much wanted to avoid mysticism so that not to irritate the scientific-minded readers any more. Nevertheless, the whole situation remains beyond our will (888):

$$\mathbf{M}_B \cdot \frac{\overset{\frown}{E}\breve{E} \cdot \mathbf{U}_{\mathbf{E}} \cdot E}{e^{5\Phi\pi e i_1\sqrt{2}}} = \frac{10^7}{\{\mathbb{N}, \mathbf{Hi}\}_{+}^{\times}}.$$

Yet, even in this improbability we have an ultimately rational picture

(889) $$\left\{\{\mathbb{N},\mathbf{Hi}\}_{+}^{\times}\cdot\chi\right\}^{\chi}=\mathbf{56099}\cdot 10^{89};$$

(890) $$\mathbf{56099}^{\oplus}=2\cdot 10^{43};$$

(891) $$\mathbf{56099}^{\otimes}=\frac{\mathbf{134365}}{9}\cdot 10^{18};$$

(892) $$\mathbf{56099}\cdot\sqrt[1+\Phi\pi e]{\mathbf{10^{90}}}=\mathbf{665}\cdot 10^{8}.$$

The mythical code numbers conserved in the ancient Greek and Mongol mathematical folklore do, indeed, always work in the unifield theory. **"My religion is very simple. I look at this Universe, so vast, so complex, so magnificent, and I say to myself that it cannot be the result of chance, but the work, however intended, of an unknown omnipotent Being, as superior to man as the Universe is superior to the finest machines of human invention."** (Napoleon)

My years long dream is to derive something very like to

(893) $$\sqrt[\oplus]{\begin{Bmatrix} X\cdot \\ \{1,2,3,4,5,6,7,8,9\}_{+}^{\times}\cdot \\ \cdot\sqrt[1+\Phi\pi e]{\mathbf{10^{90}}}\cdot \\ \cdot e^{5\Phi\pi e i_1\sqrt{2}}\cdot e^{\mathbb{N}\Phi\pi i_1} \end{Bmatrix}^{\otimes}}.$$

And, by Providence,

(894) $$\sqrt[\oplus]{\begin{Bmatrix} \mathbf{56099}\cdot \\ \{1,2,3,4,5,6,7,8,9\}_{+}^{\times}\cdot \\ \cdot\sqrt[1+\Phi\pi e]{\mathbf{10^{90}}}\cdot \\ \cdot e^{5\Phi\pi e i_1\sqrt{2}}\cdot e^{9\Phi\pi i_1} \end{Bmatrix}^{\otimes}}=\mathbf{65816}\cdot 10^{44.9999997720...}.$$

It is obviously

(895) $$\mathbf{65816=5+17+257+65537},$$

which I would like to interpret as the Hole Trinity's encoded signature.

The self-perturbation effect affecting the symmetries of the previous totality of machinery equals to

(896) $$\frac{1}{\Psi=1905494.059460...}.$$

It is easy to find that

(897)
$$\Psi\cdot\pi\frac{\Phi^3\sqrt{i_1 i_2}}{\text{Spin}}\cdot\dim_\Phi\Phi^3 i_1\cdot(3+1)\cdot$$
$$\cdot\mathbf{D\{2\}}_+^\times\cdot\mathbf{D\{3\}}_+^\times\cdot\mathbf{D\{4\}}_+^\times=\frac{\mathbf{10620052}}{81}\cdot 10^{81}\cdot$$

The $\Psi$ looks a fundamental violation of symmetries of the entire system. And it should have been compensated. What kind of compensation would have satisfied the universal system itself ?

Probably, or even certainly, the ultimate compensation is

(898)
$$\frac{\Psi\cdot\left\langle\left\|\frac{\text{🕸}}{\text{😐}}\right\|\cdot\mathbf{Cr}_+^\times\cdot\mathbf{22727}\right\rangle}{1+\frac{1}{\Gamma}}=\frac{10^{37.0000\ldots}}{\{\pi-3\}},$$

where

(899
$$\frac{10^{82}}{\Gamma^{\chi\cdot G\Pi}}=\cosh\cos^2 i,$$

closing the gigantic cosmoligical loop at

$$i=\sqrt{-1}\cdot$$

The Pi minus three is kind of the ultimate curvature of geometry:

(900)
$$\sqrt[\pi\text{-}3]{\mathbf{U}_{\mathbf{E}}}=a_e\cdot 10^{68.999\ldots};$$

(901)
$$\frac{\widehat{E}\breve{E}}{\widehat{\pi-3}}=180\cdot 10^{21}.$$

As for the boson and fermion sectors of matter ,

(902)
$$\mathbf{M}_B\{\Phi\cdot 90\cdot i_1\}=\Delta_{Exprm.}\cdot 10^6;$$

(903)
$$2N\{\Phi\cdot\Delta_{Exprm.}\cdot i_1\}=\frac{3}{\mathbf{289447}}\cdot 10^{11}.$$

The Higgs boson, if any, was originally imagined as something awfully universal in its mass-generating capacities, thus (904)

$$\mathbb{Q}\cdot\{m_e(0{,}511MeV)\cdot\dim m\}\cdot\pi\frac{\Phi^3\sqrt{i_1 i_2}}{\text{Spin}}\cdot\dim_\Phi\Phi^3 i_1\cdot(3+1)\cdot$$
$$\cdot\frac{\mathbf{D\{2\}}_+^\times\cdot\mathbf{D\{3\}}_+^\times\cdot\mathbf{D\{4\}}_+^\times}{\mathbf{InG}_+^\times}=\sin\Theta_W\cdot 10^{72}.$$

Quite surprisingly and very precisely

$$\left\{\frac{10000}{5.4461702178}(m_e = 9.10938188)\cdot \dim m\right\} = \qquad (905)$$

$$= \frac{\sqrt[\pi]{\mathbf{4790585}\cdot 10^{24}}}{\mathbb{Q}}.$$

Then, the electron itself should not be an exclusion, thus

$$\{\mathbb{Q}\cdot(m_e = 9.10938188)\cdot \dim m\}^{2\chi} = \frac{9}{\mathbf{127163}}\cdot 10^{70}. \qquad (906)$$

Remind that in the whole world there are only two stabilities, namely, proton and electron. With this in mind and taking the above into account we should smell something here. *Mein liebe Gott*, indeed,

$$\sqrt[e]{\{\mathbf{4790585}\times\mathbf{127163}\}^{2\Phi\pi i_1}} = \frac{\Phi^9}{\mathbf{65537}}\cdot 10^{59}. \qquad (907)$$

Consequently,

$$\overset{Stability}{\left\{\frac{\Phi^9}{\mathbf{65537}}\right\}} \to^{Instability} a_e. \qquad (908)$$

The lepton mass configuration is sublect to

$$\left\{\pi\left\{1+\frac{m_\mu}{m_e}+\frac{m_\tau}{m_e}\right\}\right\}^\pi \cdot \left\{\left\{\frac{\dim m}{2}\right\}^2\right\}^2 = 10^{13} \qquad (909)$$

and the Q-operator does necessarily the job as below:

$$\left\{\frac{m_\tau}{m_e} = \frac{10000}{2.87592}\right\}\mathbb{Q} = \sqrt[\pi]{\frac{\Phi^9}{\mathbf{65537}}\cdot 10^{31}}; \qquad (910)$$

$$\left\{\frac{m_\mu}{m_e} = \frac{1000}{4.83633166}\right\}\mathbb{Q}\cdot E = \frac{\mathbf{526505}}{3}\cdot 10^8. \qquad (911)$$

Now we have to approach the problem of quark masses more or less trusting the following table circulated online. However, the problem of quarks as a whole would need separate analysis in an extended way and, therefore, we are here limited to giving only one example of the Higgs mechanism of how quark masses arise. So, we consider the uud-structure from the mass spectrum point of view and have

(912) $$\left\{\frac{m_u + m_u + m_d}{m_e}\right\} = m_{uud}.$$

By the moment it is already elementary to guess that

(913) $$m_{uud} \cdot \frac{\Delta_{Exprm.}\Theta_W\Theta_{STR.}}{\sin\Delta_{Exprm.}\sin\Theta_W\cos 2\Theta_{STR.}} = \frac{10^8}{\dim^2 E}.$$

The uud-structure promises to be stable owing to the complete topological configuration of crystallographic symmetries

(914) $$\dim^2 E = \mathbf{Cr}_+^\times \cdot \cos 2\Theta_{STR.}.$$

Now just go va-banque and obtain a consistent, and inevitable, and sufficient equation with respect to the proton

(915) $$\left\{m_{uud} \cdot \frac{\Delta_{Exprm.}\Theta_W\Theta_{STR.}}{\sin\Delta_{Exprm.}\sin\Theta_W\cos 2\Theta_{STR.}} \cdot \mathbf{U_E}\right\} \cdot \mathbb{Q} \cdot \frac{10000}{\mathbf{5.4461702178}} = \pi \frac{\Phi^3 i_1}{\text{Spin}} \cdot 10^{23.9999...}.$$

Here on the right the self-perturbation effect equals to

(916) $$\frac{1}{x = 8054.155915},$$

which is simply

(917) $$x^{\Phi\pi e i_1} = \{1+2+3+4+5+6+7+8+9\} \cdot 10^{67}.$$

Yet another just one example

(918) $$\frac{m_d}{m_e} \cdot \mathbf{U_E} \cdot \mathbb{Q} = \dim^2 E \cdot 10^{14.9999...}.$$

The logic of the system requires that we would have

(919) $$1 + \frac{1}{10}\ln\lg\left\{\frac{m_d}{m_e} \cdot \mathbf{U_E} \cdot \mathbb{Q} \cdot \frac{1}{\dim^2 E}\right\} = \left\{\frac{\Phi\pi}{4} = \frac{\text{Spin}}{\sin\Delta_\beta}\right\}.$$

It is this way that from now onward any good undergraduate students of natural sciences become enabled to compute and explain on their own everything under and above the Sun.

But, those younsters should still listen to the seasoned ones like this author, particularly, when philosophy is concerned. The real philosophy comes from life experience. When younger, don't read any philosophy; just listen to old wolves. Reading philosophy weakens the rational intellect.

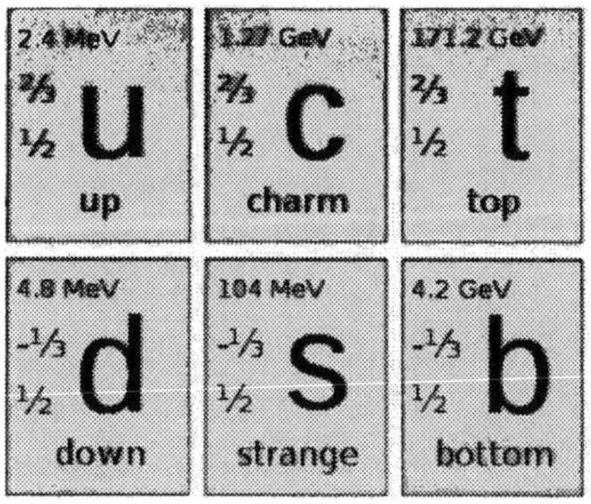


Read philosophy when retired and formulate your own which, notably, will be of real value to others.

So, what is the philosophy of the Higgs boson problem? Whether it is worth of expences at **USD10Bln** in this poor world where half of the population still suffers poverty? Physics has come to the two results: **a.** Theoretical physics reduces to the quantum gyroscope that multifurcates into the Superstructure of space-time and matter; **b.** Experimental physics reached its heights in the CERN superconducting supercollider.

What works is first of all energy fluctuation

(920) $$\left\{\left\{\sqrt[\cos 2\Theta_{STR.}]{\frac{2\Theta_{STR.}}{\cos 2\Theta_{STR.}}}\right\}^{(Fluc)}\right\}^{(Fluc)} = \mathbf{47432672396115601};$$

(921) $$\mathbf{47432672396115601}^{\chi} = \frac{3}{\mathbf{107651}}\cdot 10^{89}.$$

And it can be predicted that

(922) $$\left\{\left\{\mathbb{Q} = \mathbf{246580.2518...}\right\}^{(Fluc)}\right\}^{(Fluc)} = \mathbf{678741}\cdot 10^{10};$$

(923) $$\mathbf{678741}\cdot\pi\frac{\Phi^3\sqrt{i_1 i_2}}{\text{Spin}}\cdot\dim_{\Phi}\Phi^3 i_1\cdot(3+1)\cdot \cdot\frac{\mathbf{D\{2\}}_{+}^{\times}\cdot\mathbf{D\{3\}}_{+}^{\times}\cdot\mathbf{D\{4\}}_{+}^{\times}}{\mathbf{InG}_{+}^{\times}}\cdot\mathbf{U}_{\mathbf{E}} = \frac{10^{86}}{\mathbf{25974}};$$

(924) $$\mathbf{25974}\cdot\left\{\widehat{E}\breve{E}\cdot\mathbf{U}_{\mathbf{E}}\cdot E\right\} = \frac{\Phi}{\pi}\cdot 10^{42};$$

(925) $$\mathbf{25974}^{\chi} = ©@\cdot 10^{23}.$$

Interactions can and must be written in pure energy-entropy terms:

(926) $$\sqrt[(Fluc)]{\sqrt[(Fluc)]{\widehat{E}\breve{E}\cdot\mathbf{U}_{\mathbf{E}}\cdot E}} = e^{\Phi}\cdot 10^{12};$$

(927-928) $$\left\{\mathbb{Q}\cdot e^{\Phi}\right\}^{e^{\Phi}} = \frac{10^{35}}{\mathbf{18349}}; \qquad \left\{\mathbf{M}_{B}\cdot e^{\Phi}\right\}^{e^{\Phi}} = \Delta_0\cdot 10^{31}.$$

Now consider general configurations of nuclear physics based upon the uranium atomic number 92:

(929) $$\sqrt[\cos 2\Theta_{STR.}]{\frac{\left\{\left\{2N\cdot m_{e}\dim m\cdot e^{\pm}\dim e^{\pm}\right\}\cdot \boldsymbol{F}\times 92\right\}}{\cos 2\Theta_{STR.}}}=\frac{10^{21.9999...}}{2\pi};$$

(930) $$\left\{\left\{2N\cdot m_{e}\dim m\cdot e^{\pm}\dim e^{\pm}\cdot\Delta_{Exprm.}\right\}\cdot \boldsymbol{F}\times 92\right\}\cdot \cdot\frac{\Theta_{W}\Theta_{STR.}}{\sin\Theta_{W}\cos\Theta_{STR.}}=\sqrt[\Phi\pi]{\mathbf{\{3,5,17,257,65537\}}_{+}^{\times}\cdot 10^{52}}.$$

(931) $$\Phi\{\mathbb{Q}\cdot 2N\}=\frac{1000000000}{\mathbf{137}};$$

(932) $$\{\mathbb{Q}\cdot 2N\}^{\pi}=\mathbf{138}\cdot 10^{26}.$$

If so (933),

$$\left\{\begin{array}{l}\left\{\left\{2N\cdot m_{e}\dim m\cdot e^{\pm}\dim e^{\pm}\cdot\Delta_{Exprm.}\right\}\cdot \boldsymbol{F}\times 92\right\}\cdot\\ \cdot\dfrac{\Theta_{W}\Theta_{STR.}}{\sin\Theta_{W}\cos 2\Theta_{STR.}}\times\mathbb{Q}\end{array}\right\}^{\Phi\pi}=$$

$$=\frac{10^{95}}{\Phi\pi e}\left\{1+\frac{1}{\sin\Theta_{W}}\left\{\frac{1}{2\pi\cdot 10^{9}}\right\}^{\sin\Theta_{W}}\right\}.$$

In the very depth of reality we have a trifurcation such that

(934) $$\begin{array}{c}\cos i\\ \updownarrow\\ 2\Theta_{STR.}\leftrightarrow\ \mathbb{Q}\end{array};$$

(935) $$\{2\Theta_{STR.}\cdot\ \mathbb{Q}\}^{\cos i}\equiv\mathbf{137}\cdot 10^{9}.$$

Thus, the mass of the top quark appears to be (936, 937)

$$\sqrt[\cos 2\Theta_{STR.}]{\frac{171204.2032MeV}{m_{e}(0.511MeV)}\cdot\frac{2\Theta_{STR.}}{\cos 2\Theta_{STR.}}}=\frac{10^{22}}{\mathbf{180}};$$

$$\sqrt[\cos 2\Theta_{STR.}]{\sqrt[\cos 2\Theta_{STR.}]{\frac{171202.3111MeV}{m_{e}(0.511MeV)}\cdot\frac{2\Theta_{STR.}}{\cos 2\Theta_{STR.}}}}=\frac{10^{50}}{\cos(180-\Phi^{10})}.$$

Yet, another truly divine geometry is

(938) $$\mathbb{Q}\cdot\frac{171202.4704MeV}{m_e(0.511MeV)}=\sqrt[\exists]{\nabla\cdot10^{60}}\ .$$

In the energy-entropy terms

(939) $$\frac{171200.0031MeV}{m_e(0.511MeV)}\cdot\left\{\widehat{E}\breve{E}\cdot\mathbf{U}_{\mathbf{E}}\cdot E\right\}=\sqrt[\Phi]{\mathbf{19413}\cdot10^{65}}\ .$$

Computer simulations of the absolute geometry will provide these data automatically. **The Unifield theory is the true wealth of knowledge and freedom of learning**; it is like to what is called the embarrassment of the riche.

## Bi- and Trifurcation of Geometry

...quantum mechanics is essentially
a theory of what we do not know and
cannot predict.

**Stephen Hawking**

At first we have an algorithmic picture such that

(940-941) $$\{\Phi \cdot \arctan i_1 \cdot i_1\}^{\Phi \pi e i_1} = \frac{10^{35}}{©@};$$

$$\frac{\sqrt[e]{\{\{\Phi \cdot \arctan i_1 \cdot i_1\} \cdot \mathbf{\{3,5,17,257,65537\}_+^\times}\}^{\Phi\pi}}}{1+\sqrt[\chi]{\frac{1}{2\cdot 10^{28}}}} = \mathbf{65537} \cdot 10^{26}$$

In the second case the composition

(942) $$\{\Phi \cdot 90 \cdot i_1\}$$

images the photon

(943) $$\frac{\{\Phi \cdot 90 \cdot i_1\}^{2\Phi \pi e i_1} \cdot h \dim h}{1+\frac{1}{e^{\Phi} \cdot 10^{12}}} = 10^{81}$$

Thirdly,

(944) $$\{\Phi \cdot \Delta_{Exprm.} \cdot i_1\}^{\Phi \pi e} = \frac{\mathbf{64816}}{9} \cdot 10^{30},$$

which is spinor particles of matter. Its fundamentality is seen in

(945) $$\mathbf{64816} \cdot e^{5\Phi \pi i_1 e \sqrt{2}} = \frac{10^{\mathbf{65.9999...}}}{\mathbf{\{1,2,3,4,5,6,7,8,9\}_+^\times}};$$

(946) $$\mathbf{64816} \cdot e^{5\Phi \pi i_1 e \sqrt{2}} \left\{1+\frac{1}{z}\right\};$$

(947) $$\left\{\lg \frac{z^{G\Pi \cdot G\Pi}}{2}\right\}^{\chi} = \Phi^3 i_1 \cdot 10^8.$$

The above bi- and trifurcation process means the permanent oscillation of geometry implying that we should write a topological density in the form

(948) $$\{\arctan i_1,\ \perp,\ \Delta_{Exprm.}\}_+^\times;$$

(949): $$\left\{\{\arctan i_1,\ \perp,\ \Delta_{Exprm.}\}_+^\times \cdot \mathbf{7328}\right\}^{\Phi\pi} = \mathbf{3874}\cdot 10^{58}.$$

Now that a total configuration of this pulsating geometry will be

(950) $$\{\arctan i_1, \perp, \Delta_{Exprm.}\}_+^\times \cdot \pi \frac{\Phi^3\sqrt{i_1 i_2}}{\text{Spin}} \dim_\Phi \Phi^3 i_1 \cdot (3+1)\cdot$$
$$\cdot \mathbf{D}\{2\}_+^\times \cdot \mathbf{D}\{3\}_+^\times \cdot \mathbf{D}\{4\}_+^\times \cdot \mathbf{U}_\mathbf{E} \equiv \frac{\mathbf{10^{100}}}{\mathbf{360}};$$

$$1+W^{-1};$$

(951) $$\frac{\lg\left\{\left\{W\cdot\{\mathbf{3,5,17,257,65537}\}_+^\times\right\}^\pi \cdot \Delta_{Exprm.}\right\}}{36000} = \frac{1}{e^{\Phi\pi i_1}}.$$

(952) $$\left\{\frac{\mathbf{7328}}{\mathbf{360}}\right\}^\chi = \mathbf{4236138};$$

(953) $$\sqrt{\left\{\frac{\mathbf{3874}}{\mathbf{360}}\right\}^{\chi\chi}} = \pi\Pi_{Cosm.}\cdot 10^{13}\left\{1+\frac{1}{100\Delta_\beta}\right\}.$$

(954) $$\cos i = \sqrt{\frac{\lg \mathbf{443101}\cdot 10^{14}}{\lg\{\arctan i_1,\ \perp,\ \Delta_{Exprm.}\}_+^\times}};$$

(955) $$\mathbf{443101}^{\Phi\pi i_1} = \frac{10^{36}}{\cos 72}.$$

We have got the pure abstract geometry, that is, the soul of things. Now we shall make it with the flesh meaning physics

$$\left\{\begin{array}{l}\{\arctan i_1,\ \perp,\ \Delta_{Exprm.}\}_+^\times\cdot\\ \cdot\pi\dfrac{\Phi^3\sqrt{i_1 i_2}}{\text{Spin}}\cdot\dim_\Phi \Phi^3 i_1\cdot(3+1)\cdot\\ \cdot\mathbf{D}\{2\}_+^\times\cdot\mathbf{D}\{3\}_+^\times\cdot\mathbf{D}\{4\}_+^\times\cdot\mathbf{U}_\mathbf{E}\end{array}\right\}\times\frac{Ghm_e e^\pm c\cdot\dim(Ghm_e e^\pm c)}{\alpha a_e},$$

where the left section is equivalent to the inverse operator 1/360. Therefore,

(956) $$\left\{\frac{1}{2\pi Rad}\cdot\frac{Ghm_e e^{\pm}c\cdot\dim(Ghm_e e^{\pm}c)}{\alpha a_e}\right\}^{\chi}\cdot\frac{\mathbf{454924}}{3}\cdot 10^{41};$$

(957) $$\sqrt[\Phi]{\mathbf{454924}^{\pi e}}=\frac{10^{32.0000...}}{\Delta_1}.$$

As it is known, the ancients spoke of **Prisca Scientia**. Whether the ancients knew something real that can be proven by the latest achievements in sciences? The answer reads:

(958) $$\left\{\{\arctan i_1,\ \perp,\ \Delta_{Exprm.}\}_{+}^{\times}\times\mathbf{7328}\times\mathbf{454924}\right\}= =\sqrt[\Phi]{\mathbf{573928}\cdot 10^{23}}.$$

And

(959) $$\frac{1}{10}\ln\lg\frac{\mathbf{573928}^{G\Pi_{Cosmological}}}{\mathbf{573928}}+1=\frac{\Phi\pi}{4}=\frac{\text{Spin}}{\sin\Delta_{\beta}};$$

(960) $$\left\{\sqrt[e]{\left\{\sqrt[G\Pi]{\left\{\sqrt[\otimes]{\mathbf{7328}^{\oplus}}\right\}^{\chi}}\right\}^{\pi}}\right\}^{\Phi}=\mathbf{6719}\cdot 10^{16}.$$

The Solar system is subject to

(961) $$\odot_{+}^{\times}\cdot\mathbf{7328}=\frac{10^{49}}{2\Phi};$$

(962) $$\odot_{+}^{\times}\cdot\mathbf{7328}\cdot\mathbf{InG}_{+}^{\times}=2\Phi^{3}\cdot 10^{60.000...},$$

perturbated globally by the entire geometry

(963) $$y\cdot\pi\frac{\Phi^{3}\sqrt{i_1 i_2}}{\text{Spin}}\dim_{\Phi}\Phi^{3}i_1(3+1)\cdot \cdot\frac{\mathbf{D\{2\}_{+}^{\times}\cdot D\{3\}_{+}^{\times}\cdot D\{4\}_{+}^{\times}}}{\mathbf{InG}_{+}^{\times}}\cdot\mathbf{U_E}=\frac{10^{80}}{\Phi^{3}i_1}.$$

How it is that the mythological number is so precise a prediction?

## The Internal Geometry of Singularity (Compactification)

If you have a trigonometric circle, then you have to have

(964) $$\mathbf{InG}_{+}^{\times\exists} = \frac{10^{79}}{\mathbf{180}}.$$

It is designed by golden section algorithms and therefore

(965) $$\sqrt[\sin\Theta_W]{\mathbf{InG}_{+}^{\times}} = \frac{10^{30}}{\cos\Theta_W}.$$

Necessarily,

(966) $$\sqrt{\sqrt[\dim_\Phi \Phi^3 i_1]{\mathbf{InG}_{+}^{\times\;\pi\frac{\Phi^3 i_1}{\mathrm{Spin}}}}} = \mathbf{3345}\cdot 10^{34}\cdot$$

It is a complicated mechanism that rules the world (967)

$$\sqrt[e]{\left\{\frac{\mathbf{InG}_{+}^{\times}}{\sin\Delta_{Exprm.}\sin\Theta_W\cos\Theta_W\cos 2\Theta_{STR.}}\right\}^{\Phi\pi}} = \frac{\mathbf{263459}}{3}\cdot 10^{22}.$$

To justify all this, we find an argument such that

(968) $$\mathbf{263459}\cdot\pi\frac{\Phi^3\sqrt{i_1 i_2}}{\mathrm{Spin}}\cdot\dim_\Phi \Phi^3 i_1\cdot(3+1)\cdot$$
$$\cdot\frac{\mathbf{D\{2\}_{+}^{\times}\,D\{3\}_{+}^{\times}\,D\{4\}_{+}^{\times}}}{\mathbf{InG}_{+}^{\times}} = \frac{\mathbf{19835048}}{3}\cdot 10^{65};$$

(969) $$\mathbf{19835048}\cdot\otimes\oplus\chi = \mathbf{4276225000};$$

(970) $$\mathbf{4276225000}\times e^{5\Phi\pi i_1 e\sqrt{2}}\cdot e^{4\Phi\pi i_1} = \mathbf{69}\cdot 10^{73};$$

(971) $$\mathbf{InG}_{+}^{\times}\cdot\mathbf{U_E} = \frac{\mathbf{2650048}\cdot 10^{23}}{\mathbf{4276225000}};$$

(972) $$\mathbf{4276225000}\times\mathbf{U_E}\cdot\widehat{E}\breve{E} = \mathbf{24634}\cdot 10^{37};$$

(973) $$\left\{\frac{\mathbf{4276225000}}{E}\right\}^{\Phi\pi e i_1} = \mathbf{9141}\cdot 10^{48};$$

The inner-geometric configurations can also be

(974) $$\frac{\{\Delta_{Exprm.},\Theta_W,\Theta_{STR.}\}_+^{\times}}{\sin\Delta_{Exprm.}\,\sin\Theta_W\,\cos 2\Theta_{STR.}}=\sqrt[\Phi^3 i_1]{\Pi_{Cosm.}\cdot 10^{45}}\ ;$$

(975) $$\left\{\frac{\{\Delta_{Exprm.},\Theta_W,\Theta_{STR.}\}_+^{\times}}{\sin\Delta_{Exprm.}\,\sin\Theta_W\,\cos 2\Theta_{STR.}}\right\}^{\exists©@}=\sqrt[\pi]{\Phi\pi\cdot 10^{33}}\ ;$$

(976) $$\left\{\frac{\{\Delta_{Exprm.},\Theta_W,\Theta_{STR.}\}_+^{\times}}{\sin\Delta_{Exprm.}\,\sin\Theta_W\,\cos 2\Theta_{STR.}}\right\}\cdot$$ $$\cdot e^{5\Phi\pi i_1 e\sqrt{2}}\cdot e^{\Phi\pi i_1}\cdot\widehat{E}\breve{E}=\pi^3\cdot 10^{86}\ .$$

All roads lead, indeed, to the Rome we call human genetics:

(977) $$\ln\lg\left\{\frac{1}{5}\{\Delta_{Exprm.},\Theta_W,\Theta_{STR.}\}_+^{\times}\cdot\left[\!\left[\begin{matrix}\text{🕸}\\\text{😐}\end{matrix}\right]\!\right]\right\}=10\left\{\frac{\Phi\pi}{4}-1\right\}.$$

This mechanism is as real as anything else

(978) $$[\![\text{🕸}\text{😐}]\!]\{\Delta_{Exprm.},\Theta_W,\Theta_{STR.}\}_+^{\times}\cdot\mathbf{U_E}\times\pi\frac{\Phi^3 i_1}{\text{Spin}}=10^{43}\ .$$

There is no wonder that

(979) $$\left\{\mathbf{U_E}^{\oplus}\right\}\equiv\pi\frac{\Phi^3\sqrt{i_1 i_2}}{\text{Spin}}\ ;$$

(980) $$\pi\frac{\Phi^3\sqrt{i_1 i_2}}{\text{Spin}}\cdot\frac{\left\{\mathbf{U_E}^{\oplus}\right\}}{\mathbf{U_E}}\equiv\mathbf{1692}\ ;$$

(981) $$\frac{1}{\mathbf{65537}}\lg\left\{\frac{\left\{\mathbf{U_E}^{\oplus}\right\}}{\mathbf{U_E}}\frac{1}{\text{Spin}}\right\}=a_{Electron}\ ;$$

(982) $$\sqrt[\dim_{\Phi}\Phi^3 i_1]{\left\{\frac{\mathbf{U_E}}{a_{Electron}}\right\}^{\pi\frac{\Phi^3 i_1}{\text{Spin}}}}=\mathbf{432807}\cdot 10^{63}\ .$$

It is simply because of in-depth harmonization

(983) $$\mathbf{432807}^{\Phi\Phi\Phi\Phi}=\frac{\mathbf{385459}}{9}\cdot 10^{34},$$

where

(984) $$\Phi = 1.618\ 033\ 988\ 749...,$$

achieving the record number of true decimal places. And

(985) $$\mathbf{385459}^{\Phi\pi e i_1} = \exp e\cdot 10^{96.9999...} = \frac{100.00000...}{\mathbf{66}}.$$

As a consequence,

(986) $$\mathbf{385459}\cdot\left\{[\![\text{🕸☺}]\!]\cdot\mathbf{Cr}_{+}^{\times}\right\} = \mathbf{285}\cdot 10^{47}$$

(987) $$\frac{\left\{\mathbf{InG}_{+}^{\times}\times\left\{E\cdot\mathbf{U}_{\mathbf{E}}\cdot\widehat{E}\breve{E}\right\}\right\}^{\Phi}}{1+\delta} = \mathbf{123456789}\cdot 10^{74};$$

(988) $$\frac{1}{\delta}\times\pi\frac{\Phi^3\sqrt{i_1 i_2}}{\text{Spin}}\cdot\dim_{\Phi}\Phi^3 i_1\cdot(3+1)\cdot$$
$$\cdot\frac{\mathbf{D\{2\}}_{+}^{\times}\cdot\mathbf{D\{3\}}_{+}^{\times}\cdot\mathbf{D\{4\}}_{+}^{\times}}{\mathbf{InG}_{+}^{\times}} = \mathbf{751}\cdot 10^{78}.$$

To complete the picture (989, 990),

$$\mathbf{751}\cdot e^{5\Phi\pi i_1 e\sqrt{2}}\cdot e^{4\Phi\pi i_1} = \mathbf{121179}\cdot 10^{63};$$

$$\mathbf{121179}\cdot e^{5\Phi\pi i_1 e\sqrt{2}}\cdot e^{4\Phi\pi i_1} = \pi\frac{\Phi^3\sqrt{i_1 i_2}}{\text{Spin}}\cdot 10^{69}.$$

The logos of external and internal geometries of singularity is (991):

$$\frac{\sqrt{\pi\frac{\Phi^3 i_1}{\text{Spin}}\cdot\pi\frac{\Phi^3\sqrt{i_1 i_2}}{\text{Spin}}\cdot\dim_{\Phi}\Phi^3 i_1\cdot(3+1)\cdot}\ \cdot\mathbf{D\{2\}}_{+}^{\times}\cdot\mathbf{D\{3\}}_{+}^{\times}\cdot\mathbf{D\{4\}}_{+}^{\times}}{\mathbf{InG}_{+}^{\times}}\cdot\mathbf{U}_{\mathbf{E}} = ©\cdot 10^{76}.$$

Now that the Unified cosmologicsl field as a whole (992)

$$\frac{\sqrt{\pi\frac{\Phi^3 i_1}{\text{Spin}}\cdot\pi\frac{\Phi^3\sqrt{i_1 i_2}}{\text{Spin}}}\dim_{\Phi}\Phi^3 i_1(3+1)\mathbf{D\{2\}}_{+}^{\times}\mathbf{D\{3\}}_{+}^{\times}\mathbf{D\{4\}}_{+}^{\times}}{\mathbf{InG}_{+}^{\times}}\mathbf{U}_{\mathbf{E}}\cdot$$
$$\cdot\frac{Ghm_e e^{\pm}c\cdot\dim(Ghm_e e^{\pm}c)}{\alpha a_e} = \sqrt[\cos 2\Theta_{STR.}]{\mathbf{360}\cdot 10^{31}}.$$

A bigbang cosmological scenario

(993) $$\text{Unifield}\cdot\exists = e^{e}\cdot 10^{85}\cdot$$

The theory of mathematical continuum begins with the trinity of constants so that to end up with the intelligent life becoming able to write

(994) $$\Phi\pi e\cdot\text{Unifield}\left[\!\left[\circledast\cdot\mathbf{Cr}_{+}^{\times}\right]\!\right] = \frac{4}{\pi}\cdot 10^{122}.$$

It is simply because of

(995) $$\sqrt[\sin\Delta_{Exprm.}]{\frac{4}{\pi}} = \sqrt[\Delta_{Exprm.}]{\mathbf{124201}\cdot 10^{6}}\ ;$$

(996) $$\mathbf{124201}\cdot e^{10\Phi\pi i_1} e^{5\Phi\pi i_1 e\sqrt{2}} = \mathbf{1414594}\cdot 10^{81.0000000000\ldots}.$$

Having achieved its goal, geometry makes self-reference

(997) $$\mathbf{1414594}\times\Phi\pi e = \pi\frac{\Phi^{3} i_1}{\text{Spin}}\cdot 10^{6}.$$

It is how the old mathematics gives birth to the new:

(998) $$1000\ 000\cos^{2} i \equiv \pi\frac{\Phi^{3}\sqrt{i_1 i_2}}{\text{Spin}}\cdot\{\Delta_{Exprm.}\Theta_{W}\Theta_{STR.}\}$$

and accurately and remarkably

(999) $$1000\ 000\cos^{2} i = \frac{\pi\dfrac{\Phi^{3}\sqrt{i_1 i_2}}{\text{Spin}}\cdot\{\Delta_{Exprm.}\Theta_{W}\Theta_{STR.}\}}{1-\left\{\dfrac{1}{\mathbf{InG}_{+}^{\times}\cdot 10^{9}}\right\}^{\Phi\pi i_1}}.$$

My article on the Superstructure and related matters was deposited in the archives of **Physical review/D** in 1999:

**Date:** Mon, 25 Oct 1999 14:39:45 -0400 (EDT)
**From:** PR/D-email <prd@aps.org> | Block address
**To:** erdenee_ch@yahoo.com
**Subject:** RE: DK7178
**CC:** prd@aps.org

DK7178
Principia Mathematica I
Erdeni, Chu/

## Sidereal Year

The unifield theory discovers that the Solar system of planets is too much perfect a structure from the absolute geometric point of view. In this situation we can no more rule out the possibility of artificial origin of our greater home.

Then, whether the sidereal year of the Earth has a meaning or not? We refer to Wikipedia:

"A sidereal year is the time taken by the Earth to orbit the Sun once with respect to the fixed stars... It was equal to 365.256363004 SI days at noon 1 January 2000.".

It is, indeed, a code number

(1000) $$\{\mathbf{SY}\}^{2\Phi\pi i_1 e} = \frac{10^{88.9999...}}{\Phi\cdot(©@)^2};$$

(1001) $$\sqrt[\sin i\cos i]{\{\mathbf{SY}\}\cdot e^{5\Phi\pi i_1 e\sqrt{2}}} = \sqrt{©@\cdot 10^{63}}.$$

It is much more powerful than we can even imagine, for (1002)

$$\{\mathbf{SY}\}\cdot\frac{\dfrac{\left\{\begin{matrix}\dfrac{Ghm_e e^{\pm}c\cdot\dim(Ghm_e e^{\pm}c)}{\alpha a_e}\cdot\Theta_W\Theta_{STR.}\cdot\\ \cdot\{\{2\times 2\cdot 6\}\cdot \boldsymbol{FB}\}\end{matrix}\right\}}{\sin\Delta_{Exprm.}\sin\Theta_W\cos\Theta_W\cos 2\Theta_{STR.}} = \mathbf{2.920925121}\cdot 10^{15}}{\mathbf{92}} = a_e\cdot 10^{19.00000...}.$$

Therefore, in general,

(1003) $$\{\mathbf{SY}\}\cdot\left\{\left\{\odot_{+}^{\times}\left[\left[\right.\right.\right.\right.$$  $$\left.\left.\left.\right]\right]\right\}\{\mathbf{Cr}_{+}^{\times}\cdot\mathbf{22727}\}\right\}\equiv\exists.$$

## Mass Spectra

The latest experimental data on the masses of quarks

(1004) $$\frac{\left\{\begin{matrix}2.4+1270+171200+\\+4.8+104+4200\end{matrix}\right\}MeV}{m_e(0.511MeV)}=\mathbf{M_Q};$$

(1005) $$\mathbf{M_Q}\cdot Gh\cdot\left\{\dim Gh=\Phi^3 i_1\right\}=\mathbf{82423107}.$$

(1006) $$\mathbf{M_Q}\cdot e^{2\Phi\pi i_1}=\mathbf{1429733}\cdot 10^5;$$

(1007) $$\mathbf{M_Q}\cdot\Delta_{Exprm.}=\mathbf{47407805};$$

(1008) $$\mathbf{M_Q}=\sqrt[\chi]{\frac{10^{28}}{\cos\Theta_W}};$$

(1009) $$\mathbf{M_Q}\cdot\frac{\Theta_W\Theta_{STR.}}{\sin\Theta_W\cos\Theta_W\cos 2\Theta_{STR.}}\equiv\pi\frac{\Phi^3\sqrt{i_1 i_2}}{\text{Spin}};$$

(1010) $$\mathbf{M_Q}\cdot e^{5\Phi\pi e i_1\sqrt{2}}\cdot\left\{\widehat{E}\breve{E}\cdot\mathbf{U_E}\cdot E\right\}\equiv\frac{1}{\cos i};$$

(1011) $$\mathbf{M_Q}\cdot\pi\frac{\Phi^3\sqrt{i_1 i_2}}{\text{Spin}}\cdot\dim_\Phi\Phi^3 i_1\cdot(3+1)\equiv\tan^2\frac{2\pi}{5}.$$

How about the Q-operator ?

(1012) $$\mathbb{Q}\cdot\mathbf{M_Q}=\mathbf{85304800000}.$$

We compute with

(1013, 1014) $$\mathbf{M_Q}=345951.4677;$$ $$\mathbb{Q}=246580.2518.$$

With respect to the boson masses we have before had

(1015) $$\frac{126002.5087MeV+\\+\left\{2\times 80398MeV+91187.6MeV\right\}}{0,511MeV}=\\=\mathbf{M_B}=739698.8429.$$

We work with the following characteristic exactitude

(1016) $$\mathbf{M}_B\mathbf{M}_\mathrm{Q} \cdot \mathbf{U}_\mathrm{E} = \mathbf{5784} \cdot 10^{16.9999999...}.$$

Now see a remarkable

(1017) $$e^{5\Phi\pi e i_1 \sqrt{2}} \cdot \{\mathbb{Q} \cdot \mathbf{M}_B\mathbf{M}_\mathrm{Q}\} \cdot \{\widehat{E}\breve{E} \cdot \mathbf{U}_\mathrm{E} \cdot E\} = \frac{10^{103.000...}}{\alpha a_e}.$$

The lepton mass configuration is

(1018) $$1 + \left\{\frac{m_\tau}{m_e} = \frac{10000}{2.87592}\right\} + \left\{\frac{m_\mu}{m_e} = \frac{1000}{4.83633166}\right\} = \\ = \mathbf{L}_\mathrm{L} = 3684.916467.$$

So,

(1019) $$\{\mathbb{Q} \cdot \{\{\mathbf{M}_\mathrm{Q}\mathbf{L}_\mathrm{L}\} \cdot \mathbf{M}_B\}\} \cdot \frac{\Delta_{Exprm.}}{\sin \Delta_{Exprm.}} = \mathbf{46752} \cdot 10^{18};$$

(1020) $$\{\mathbb{Q} \cdot \{\{\mathbf{M}_\mathrm{Q}\mathbf{L}_\mathrm{L}\} \cdot \mathbf{M}_B\}\} \cdot e^{2\Phi\pi i_1} = \mathbf{960939} \cdot 10^{20}.$$

Any problem is in the end continuum-theoretical one, therefore,

(1021) $$\{\mathbb{Q} \cdot \{\{\mathbf{M}_\mathrm{Q}\mathbf{L}_\mathrm{L}\} \cdot \mathbf{M}_B\}\} \cdot e^{5\Phi\pi e i_1 \sqrt{2}} = \frac{\mathbf{659}}{3} \cdot 10^{72};$$

(1022) $$\{\mathbb{Q} \cdot \{\{\mathbf{M}_\mathrm{Q}\mathbf{L}_\mathrm{L}\} \cdot \mathbf{M}_B\}\} \cdot \pi \frac{\Phi^3 \sqrt{i_1 i_2}}{\mathrm{Spin}} \cdot \dim_\Phi \Phi^3 i_1 \cdot (3+1) \cdot \\ \cdot \frac{\mathbf{D}\{2\}_+^\times \cdot \mathbf{D}\{3\}_+^\times \cdot \mathbf{D}\{4\}_+^\times}{\mathbf{InG}_+^\times} \cdot \left\{\left\{\frac{\dim E}{2}\right\}^2\right\}^2 = 10^{87}.$$

Now that a superunified field equation shall look

(1023) $$\frac{\frac{2N \cdot Ghm_e e^{\pm} c \cdot \dim(...)}{\alpha a_e} \Theta_W \Theta_{STR.} \{24 \cdot \boldsymbol{FB}\} \cdot \\ \cdot \{\mathbb{Q}\{\{\mathbf{M}_\mathrm{Q}\mathbf{L}_\mathrm{L}\}\mathbf{M}_B\}\} \cdot \{\widehat{E}\breve{E} \cdot \mathbf{U}_\mathrm{E} \cdot E\} \equiv \mathbf{53} \cdot 10^{74}}{e^{5\Phi\pi e i_1 \sqrt{2}}} = \\ = \mathbf{561} \cdot 10^{18.999999...}.$$

## The Cosmic Superfield
## &
## Transcosmic Information Circulation

To compose superfield is now elementary; important is just to know principal sections of geometry and physics. For example (1015),

$$\mathbf{S} = \frac{\exists}{©@} \times \frac{Ghm_e e^{\pm} c \cdot \dim(Ghm_e e^{\pm} c)}{\alpha a_e} \cdot$$

$$\cdot \frac{\Theta_W \Theta_{STR.}}{\sin \Delta_{Exprm.} \sin \Theta_W \cos \Theta_W \cos 2\Theta_{STR.}} \cdot \left\{ \{2 \times 2 \cdot \mathbf{6}\} \cdot \boldsymbol{FB} \right\} \times \mathbf{92}.$$

Then,

(1016) $$\sqrt[\dim_{\Phi} \Phi^3 i_1 \cdot (3+1)]{\{\mathbf{S}\}^{\pi \frac{\Phi^3 i_1}{\text{Spin}}}} \times \mathbf{D}\{4\}_{+}^{\times} = \frac{10^{74}}{\mathbf{180}}.$$

Transcosmic information flow in the Universe is, in fact, inevitable a component of the superfield medium:

(1017) $$\frac{\{\mathbf{1292049,\ 310952}\}_{+}^{\times}}{\mathbf{180}} = \sqrt[\pi]{\frac{10^{51}}{\Delta_{Exprm.}}}.$$

(1018) $$\frac{\left\{ \lg \left\{ \frac{1}{\nabla} \cdot \sqrt[\Phi]{\frac{10^{74}}{\mathbf{180}}} \times \{\mathbf{1292049,\ 310952}\}_{+}^{\times} \right\} \right\}^{\chi}}{1 + \frac{e^{\Phi\pi}}{10^7}} = \Phi^3 i_1 \cdot 10^8.$$

(1019) $$e^{5\Phi\pi i_1 e\sqrt{2}} \cdot \frac{\mathbf{180}}{\{\mathbf{1292049,\ 310952}\}_{+}^{\times}} = \frac{\mathbf{7921283}}{3} \cdot 10^{32},$$

And necessarily

(1020) $$\sqrt[\dim_{\Phi} \Phi^3 i_1 \cdot (3+1)]{\mathbf{7921283}^{\pi \frac{\Phi^3 i_1}{\text{Spin}}}} = \mathbf{4286523900}.$$

## Theta-Unified Physics

So great is the power of truth.
**Kepler**

In the unifield theory it is proved to be (1021)

$$\Theta_{Electroweak} = \arctan\frac{1}{2}: \quad \frac{\pi}{3} - \Theta_W = \Theta_{Nuclear\ Strong\ Force}\ .$$

It is so, not otherwise, simply because of

(1022) $$\frac{\{3,5,17,257,65537\}_+^\times}{\Theta_W\Theta_{STR.}} = \frac{10^{11.9999...}}{\pi}\ .$$

If accuracy needed, then

(1024) $$\sqrt[\dim_\Phi \Phi^3 i_1]{\left\{\frac{1}{\delta}\right\}^{\Phi^3 i_1}} = \frac{\mathbf{3443569}}{3}\ .$$

Interpret the Theta-Alpha-Aleph origin of the gene code

(1025) $$\{\circledast\cdot\Theta_W\Theta_{STR.}\}\times\mathbf{3} = \frac{10^{15}}{\alpha a_e}\ .$$

Now memorize that the Solar system of planets is a living organism as precisely fine tuned the underlying parameters as

(1026) $$\Theta_W\Theta_{STR.}\times\left\{\odot_+^\times\cdot\left[\!\left[\frac{\circledast}{\circleddash}\right]\!\right]\right\} = \mathbf{78031}\cdot 10^{51}\ .$$

A theorem on the anthropic evolution reads

(1027) $$[\![\circledast\circleddash]\!] = \sqrt[\cos\tilde{i}]{\frac{\mathbf{14513248}}{3}\cdot 10^{23}}\ ;$$

(1028) $$\frac{\mathbf{154836500000000}}{9} = \mathbf{14513248}\times\sqrt[\Phi\pi e+1]{10^{90}}\ ,$$

(1029) $$\mathbf{1548365}^{\Phi\Phi\Phi\Phi} = \frac{\mathbf{8001001}}{3}\cdot 10^{36}\ ,$$

where

$$\Phi = \mathbf{1.618\,033\,988\,749\,9}\ .$$

The above lasts longer and ends up with

(1030) $$\sqrt[©]{\mathbf{6217195}} = \frac{\mathbf{8567684000000}}{9};$$

(1031) $$\mathbf{6217195}^{\exists} = \frac{\mathbf{56558273}}{9} \cdot 10^{32}.$$

The nuclear strong force parameter satisfies

(1032) $$e^{\Phi\Theta_{\textit{Nuclear Strong Force}}} = \frac{10^{30}}{\mathbf{3200025}}.$$

Therefore, the **Theta-strong unification of physice** will be

(1033) $$\left\{\frac{\exists}{©@}\left\{\frac{Ghm_e e^{\pm}c \cdot \dim(Ghm_e e^{\pm}c)}{\alpha a_e}\right\}\Theta_W \Theta_{STR.}\right\}\cdot$$
$$\cdot \boldsymbol{FB} \equiv e^{\Theta_{\textit{Nuclear Strong Force}}},$$

where the to self-perturbation effect of geometry is defined by

(1034) $$\left\{\frac{1}{x}\right\}^{\Phi\pi e i_1} \times \pi \frac{\Phi^3 i_1}{\text{Spin}} = 10^{38}.$$

## A Sample of Operators

In calculations occurs an operator which is

(1035) $$\mathfrak{I} = \cos\tilde{\mathbf{i}} \cdot \sqrt{i_1 i_2} \cdot \left\{\left\{\frac{\dim E}{2}\right\}^2\right\}^2 \cdot \boldsymbol{FB} = \mathbf{8.9195912...}$$

and it may lead to compositions like

(1036) $$\frac{\exists}{©@}\pi\frac{\Phi^3\sqrt{i_1 i_2}}{\text{Spin}}\dim_{\Phi}\Phi^3 i_1(3+1) = \sqrt[\exists]{\frac{10^{34}}{©}};$$

(1037) $$\mathfrak{I}\cdot\pi\frac{\Phi^3\sqrt{i_1 i_2}}{\text{Spin}}\frac{\dim_{\Phi}\Phi^3 i_1(3+1)}{\alpha a_e} = \frac{\mathbf{865735700}}{3}.$$

## An Alien Message Decyphered
## or
## The Intergalaxy IQ-test

Mysticism has not the patience
to wait for God's revelation.
***Kierkegaard***

You cannot really understand any
myths till you have found out that
one of them is not a myth.
***Chesterton***

One of the Russian research institutions should be possessing a metal fragment from the outer Cosmos, if, of course, the piece survived the notoriously known privatization process.

The fragment is said to have been found near the river Vakhsha in 1976. What is of much interest to us, it was an alloy of rare earth elements of the following composition:

Cerium - 67.2 %;
Lantanum - 10.9 %;
Neodymum - 8.78 %.

We have two overlapping topological configurations:

(1048) $$\left\{\mathbf{Ce}^{58}_{67.2},\ \mathbf{La}^{57}_{10.9},\ \mathbf{Nd}^{60}_{8.78}\right\}^{\times}_{+}.$$

So, first (1049)

$$\left\{\begin{array}{l}\left\{\left[\{\mathbf{67.2,\ 10.9,\ 8.78}\}^{\times}_{+}\cdot X\right]^{\pi}=\Delta_{\alpha}\cdot 10^{35}\right\}^{\Phi}= \\ \qquad\qquad =\Phi\pi e\cdot\Theta_{W}\Theta_{STR.}\cdot 10^{56}\end{array}\right\}^{i_1}=e\cdot 10^{\mathbf{76}}.$$

(1050) $$\sqrt[3]{\{\Phi\pi e\}^{76}}=\mathbf{777936}\cdot 10^{23};$$

(1051) $$\mathbf{777936}\cdot\left\{[\![❁☺]\!]\cdot\mathbf{Cr}^{\times}_{+}\right\}=\frac{\mathbf{17255618}}{3}\cdot 10^{180-137};$$

(1052) $$\mathbf{17255618}\cdot\left\{\odot^{\times}_{+}\cdot 360\right\}=\frac{\mathbf{23582}}{9}\cdot 10^{51}.$$

After all, it is simply that

(1053) $$\{\mathbf{67.2,\ 10.9,\ 8.78}\}^{\times}_{+}\cdot ❁=\frac{\mathbf{7433654}}{3}\cdot 10^{16}$$

Now we will deal with

(1054)
$$\left\{\mathbf{Ce}_{67.2}^{58},\ \mathbf{La}_{10.9}^{57},\ \mathbf{Nd}_{8.78}^{60}\right\}_{+}^{\times}= =\left\{\mathbf{67.2,\ 10.9,\ 8.78}\right\}_{+}^{\times}\cdot\left\{\mathbf{58,57,60}\right\}_{+}^{\times}.$$

So, geometry's F-invariant self-reference is

(1055)
$$\sin\lg\frac{\left\{\mathbf{Ce}_{67.2}^{58},\ \mathbf{La}_{10.9}^{57},\ \mathbf{Nd}_{8.78}^{60}\right\}_{+}^{\times\chi}}{\left\{\mathbf{Ce}_{67.2}^{58},\ \mathbf{La}_{10.9}^{57},\ \mathbf{Nd}_{8.78}^{60}\right\}_{+}^{\times}}=\frac{\Phi}{2};$$

(1056)
$$\left\{\mathbf{Ce}_{67.2}^{58},\ \mathbf{La}_{10.9}^{57},\ \mathbf{Nd}_{8.78}^{60}\right\}_{+}^{\times\Phi}=\frac{\mathbf{28457476}}{9}\cdot10^{15}.$$

(1057) $$\mathbf{28457476\times\{3,5,17,257,65537\}}_{+}^{\times}=\sqrt{2\Phi^{3}\cdot10^{43}}.$$

The cosmic brothers confirm that we are not wrong in writing the trinity of energy and entropy

(1058) $$\left\{\mathbf{Ce}_{67.2}^{58},\ \mathbf{La}_{10.9}^{57},\ \mathbf{Nd}_{8.78}^{60}\right\}_{+}^{\times}\times\left\{\mathbf{U}_{\mathrm{E}}\cdot\hat{E}\breve{E}\cdot E\right\}=\frac{10^{52}}{\mathbf{26}}.$$

The alien message also provides

(1059)
$$\mathbf{1528}^{\Phi\pi e i_{1}}=\mathbf{92235}\cdot10^{51}$$

It is somewhat foreseeable that

(1060)
$$\sqrt[\sin137]{\left\{\mathbf{Ce}_{67.2}^{58},\ \mathbf{La}_{10.9}^{57},\ \mathbf{Nd}_{8.78}^{60}\right\}_{+}^{\times}\times\frac{137}{\sin137}}= =\mathbf{72482}\cdot10^{18}.$$

And

(1061)
$$\sqrt[\cos2\Theta_{STR.}]{\frac{\mathbf{72482}}{\cos2\Theta_{STR.}}}=\frac{10^{12.9999...}}{\cos2\Theta_{STR.}}.$$

Moreover,

(1062)
$$\left\{\mathbf{Ce}_{67.2}^{58},\ \mathbf{La}_{10.9}^{57},\ \mathbf{Nd}_{8.78}^{60}\right\}_{+}^{\times}\cdot\mathbf{Cr}_{+}^{\times}\cdot \cdot\left\{\mathbf{U}_{\mathrm{E}}\cdot E\right\}=\left\{\Delta_{Exprm.}\aleph_{1}=\frac{1}{\alpha a_{e}}\right\}\cdot10^{42}.$$

The Maya's cycle of 5126 years ended in 2012. We have (1063)

$$\{\mathbf{5126}\times 365.256\}\cdot\pi\frac{\Phi^3\sqrt{i_1 i_2}}{\text{Spin}}\cdot\dim_\Phi\Phi^3 i_1\cdot(3+1)\cdot$$

$$\cdot\frac{\mathbf{D\{2\}_+^\times\cdot D\{3\}_+^\times\cdot D\{4\}_+^\times}}{\mathbf{InG_+^\times}}\cdot\mathbf{U_E}\cdot\odot_+^\times=\frac{\mathbf{403163}}{9}\cdot 10^{40}.$$

and subsequently and naturally

(1064) $$\mathbf{403163}\cdot\{\circledast\mathbf{Cr}_+^\times\}=\mathbf{14}\cdot 10^{39.99999...};$$

(1065) $$\frac{\mathbf{403163}}{\mathbf{9}}\cdot\{\circledast\mathbf{Cr}_+^\times\}=\frac{10^{43.00000...}}{e^{\Phi\pi i_1}}.$$

The Mayan cycle is an energy-related phenomenon:

(1066) $$\mathbf{403163}\times E=\frac{\mathbf{1249}}{9}\cdot 10^9;$$

(1067) $$\sqrt[\cos 2\Theta_{STR.}]{\mathbf{403163}\frac{2\Theta_{STR.}}{\cos 2\Theta_{STR.}}}=\mathbf{89}\cdot 10^{18}.$$

Therefore, immediately

(1068) $$\mathbf{89}\cdot\circledast\cdot\mathbf{Cr}_+^\times=\cos 72\cdot 10^{38};$$

(1069) $$\mathbf{89}\cdot\{[\![\circledast\circleddash]\!]\cdot\mathbf{Cr}_+^\times\}\cdot\mathbf{22727}=\frac{10^{52}}{2\Theta_{STR.}}.$$

In the unifield theory it is a must to write the origin of time as

(1070) $$\Phi\cdot\arctan i_1\cdot i_1,$$

and the aliens would hardly miss this problem. Indeed,

(1071) $$\{\mathbf{Ce_{67.2}^{58},\ La_{10.9}^{57},\ Nd_{8.78}^{60}}\}_+^\times\cdot\{\Phi\cdot\arctan i_1\cdot i_1\}=$$
$$=\mathbf{2068913}\cdot 10^9.$$

And magically

(1072) $$\mathbf{2068913}\cdot e^{5\Phi\pi i_1 e\sqrt{2}}=\pi\frac{\Phi^3 i_1}{\text{Spin}}\cdot 10^{59};$$

(1073) $$\mathbf{365.256}\cdot\frac{\Phi^3\sqrt{i_1 i_2}}{\text{Spin}}\cdot\dim_\Phi\Phi^3 i_1\cdot(3+1)\equiv\mathbf{100000}.$$

## What is Synthesis ?

A theorem on the algorithmic origin of Existence (1074) :

$$\sqrt[e]{\left\{ \mathbf{\{3,5,17,257,65537\}}_{+}^{\times} \cdot \frac{G}{\Pi} \right\}^{2\Phi\pi i_1}} = \frac{\mathbf{2638453}}{3} \cdot 10^{68}$$

has its immediate self-proof

(1075) $$\mathbf{2638453}^{G\Pi} = \mathbf{1401511} \cdot 10^{17} \cdot$$

The relativistic interval looks somewhat strange because of the signature problem

(1076) $$ds^2 = c^2dt^2 - dx^2 - dy^2 - dz^2 .$$

To correct the situation, the 4th coordinate is usually written as

(1077) $$cit \cdot$$

The Poisson brackets in quantum mechanics appear in the form

(1078) $$[z, w] = \frac{i}{h}(zw - wz).$$

Then, why not to write from the beginning

(1079) $$x + yi = \Phi + ii_1 \ ?$$

In this case the move to synthesis is as instantaneous as

(1080) $$\sqrt[\chi]{e^{\Theta_W \cdot \Phi\pi i_1}} = \Phi^3 i_1 \cdot 10^{14} .$$

The golden algorithm is the thoroughgoing process

(1081) $$\sqrt[\sin\Theta_{Weinberg(Electroweak)}]{\Delta_{Exprm.}\aleph_1} = \frac{9}{\mathbf{409046}} \cdot 10^{16} ;$$

(1082) $$\mathbf{409046}^{\Phi\pi i_1} = \mathbf{1929096} \cdot 10^{30} \cdot$$

The next step is evident:

(1083) $$\sqrt[\cos 2\Theta_{Strong\ Nuclear\ Force}]{\Delta_{Exprm.}\aleph_1} = \frac{10^{17.99988...}}{\Delta_1\Theta_W\Theta_{STR.}} ,$$

where the perturbation effect is defined by

(1084) $$1 - \frac{9}{\mathbf{33895}} ;$$

(1085) $$\mathbf{33895} \cdot \pi \frac{\Phi^3\sqrt{i_1 i_2}}{\text{Spin}} \cdot \dim_\Phi \Phi^3 i_1 \cdot (3+1) = \mathbf{9280000} ;$$

In fact the only risky postulate in this geometry is that of the strong force parameter. Although it always justifies itself, it is still important to know that

(1086) $$\sqrt[\chi]{e^{\Theta_{STR.}\cdot\Phi\pi i_1}} = \frac{3}{\mathbf{863}}\cdot 10^{21};$$

(1087) $$\mathbf{863\times D\{2\}_+^\times \cdot D\{3\}_+^\times \cdot D\{4\}_+^\times} = \frac{\mathbf{6506597}}{3}\cdot 10^{74}.$$

The following proves that inside the superstructure does, indeed, exist algorithmically constructable Theta angles:

(1089) $$e^{\pi \arctan i_1} = \frac{\Phi}{\pi} = \frac{\exists}{\mathbf{D\{4\}_+^\times}};$$

(1090) $$e^{\Phi\pi\cdot\arctan i_1} = \frac{10^{93}}{i_1\Theta_W\Theta_{STR.}};$$

(1091) $$\frac{\left\{\arctan i_1\cdot 90\cdot\Delta_{Exprm.}\right\}}{a_e} = \mathbf{551198442},$$

whence

(1092) $$\frac{1}{\alpha a_e} = \mathbf{118169.9226974351554494957379 05}$$

implying that the two anomalies are defined intrinsically together. If, say,

$$\alpha = \frac{1}{137.035999},$$

then

(1093) $$a_e = 0.00115965209989067994345811173697 36.$$

The kinematic angles of geometry come from fundamental principles of constructive algorithms (1094-1095)

$$\sqrt[e]{\left\{\arctan i_1\cdot 90\cdot\Delta_{Exprm.}\cdot\Theta_W\cdot\Theta_{STR.}\times\mathbf{U_E}\right\}^{\Phi\pi i_1}} = \Delta_1\frac{65537}{\Phi^9}\cdot 10^{38};$$

$$\{\arctan i_1, 90, \Delta_{Exprm.}, \Theta_W, \Theta_{STR.}\}_+^\times\cdot e^{3\Phi\pi i_1}\times\pi\frac{\Phi^3\sqrt{i_1 i_2}}{\text{Spin}} = 10^{21}.$$

# Superunification Solution for Physics in the Theory of Mathematical Continuum

Galileo was a great genius, and so was Newton; but it would take two or three Galileos and Newtons to make one Kepler.

**Samuel Taylor Coleridge**

The mathematical system of the universal harmony is, finally, such that (1096)

$$\left\{\frac{\exists}{\copyright @}X\left\{\frac{\frac{2N\cdot Ghm_e e^{\pm}c\cdot\dim(Ghm_e e^{\pm}c)}{\alpha a_e}\cdot\Theta_W\Theta_{STR.}\cdot\{2\times 2\cdot\mathbf{6}\cdot\boldsymbol{FB}\}\times\mathbf{92}}{\sin\Delta_{Exprm.}\sin\Theta_W\cos\Theta_W\cos 2\Theta_{STR.}}\right\}\equiv\frac{\mathbf{265885}}{9}\right\}\cdot$$

$$\cdot\pi\frac{\Phi^3\sqrt{i_1 i_2}}{\text{Spin}}\dim_\Phi\Phi^3 i_1(3+1)\cdot$$

$$\cdot\frac{\mathbf{D\{2\}}_+^\times\cdot\mathbf{D\{3\}}_+^\times\cdot\mathbf{D\{4\}}_+^\times}{\mathbf{InG}_+^\times}\cdot\cdot\left\{\hat{E}\breve{E}\cdot\mathbf{U}_{\mathrm{E}}\cdot E\right\}\cdot$$

$$\cdot e^{5\Phi\pi e i_1\sqrt{2}}=\frac{10^{\mathbf{188}.00000...}}{\mathbf{72}}.$$

If accurately,

$$\text{(1097)}\qquad ...=\frac{10^{\mathbf{188}}}{\mathbf{72}}1+\sqrt[e]{\left\{\frac{2}{\Phi\cdot 10^{24}}\right\}^{2\Phi\pi i_1}}.$$

The universal Φ-invariance (or fundamental metric) and the symmetry of the fifth order are equivalent concepts in absolute geometry. That's why the superunified description of the world should look as above, not otherwise.

Since geometry insists on its fundamental symmetry, we will follow the pattern and unify the problem of the existence of observers with physics as simply as next

(1098) $$\frac{\text{☺}}{72^2} = \frac{369656}{9};$$

(1099) $$\frac{\text{⛤}}{72^6} \equiv \frac{1000\ 000}{\pi};$$

(1100) $$\sqrt[\Phi]{\mathbf{\{Genes\}}_{+}^{\times}} = \tan 72 \cdot 10^{45.99999...}$$

(1101) $$\frac{\odot_{+}^{\times}}{\tan 72} = \Delta_{\beta} \cdot 10^{42}.$$

The origin of the pentasymmetry of the Superstructure (which is an oblate pentagon) lies in

(1102) $$\sqrt[1+\Phi\pi e]{\mathbf{10^{90}}} \cdot e^{\Phi\pi i_1} e^{5\Phi\pi i_1 e\sqrt{2}} = \frac{2\pi}{5} \cdot 10^{60.9999...}.$$

The fundamental pentasymmetry of Nature is broken by the standard self-perturbation of the entire geometry

(1103) $$\frac{1}{\varepsilon} \cdot \pi \frac{\Phi^3 \sqrt{i_1 i_2}}{\text{Spin}} \dim_{\Phi} \Phi^3 i_1 (3+1) \cdot \\ \cdot \frac{\mathbf{D\{2\}_{+}^{\times} \cdot D\{3\}_{+}^{\times} \cdot D\{4\}_{+}^{\times}}}{\mathbf{InG_{+}^{\times}}} \cdot \mathbf{U_E} = \mathbf{64} \cdot 10^{77}.$$

In most major equations do matter even powers of ten on the right. In the given case

(1104) $$\pi^{\mathbf{188}} = 180\Phi \cdot 10^{91}\cdot$$

The numbers 23, 46,and 92 signify the end of natural evolution:

(1105) $$\mathbf{\{23,46,92\}}_{+}^{\times\, G\Pi_{Cosmo-Logical}} = \aleph_1 \cdot 10^{23};$$

(1106) $$\mathbf{\{23,46,92\}}_{+}^{\times} X \cdot e^{5\Phi\pi e i_1 \sqrt{2}} \cdot \widehat{E}\breve{E} = \sin\Theta_W \cdot 10^{90};$$

(1107) $$\left\{\frac{\mathbf{\{23,46,92\}}_{+}^{\times}}{72}\right\}^{\chi \cdot G\Pi} = \frac{\mathbf{2703107}}{9} \cdot 10^{91}.$$

(1108) $$\ln \lg \frac{\sqrt[e]{\mathbf{2703107}^{\Phi\pi i_1}}}{2} \Rightarrow \{\boldsymbol{F} \leftrightarrow \boldsymbol{B}\}.$$

Therefore, it is right at the moment when one writes and another reads this text the natural evolution does around up achieving its final goal. The Universal machine, this Supreme Intelligence, proves that its self-modelling experiment in the capacity of human brain is going to succeed eventually. It is not that we are so good to discover this geometry. The Cosmos cares about the fate of its unique experiment. The Universal Mathematical Machine reproduces itself in the human brain and in our fundamental knowledge at any given historical moment.

Every one of us once began to count our fingers. The modern civilization returns back to this simplest but most efficient method: We will compute things and phenomena in Nature in terms of natural numbers said to be God's only creation. For example,

(1109) $$\mathbf{663176}\frac{e^{5\Phi\pi i_1 e\sqrt{2}}}{\widehat{E}\breve{E}} = \frac{\mathbf{7374655}}{3}\cdot 10^{31};$$

(1110) $$\sqrt[\dim_{\Phi}\Phi^3 i_1]{\mathbf{7374655}^{\pi\frac{\Phi^3 i_1}{\mathrm{Spin}}}} = \mathbf{24646}\cdot 10^{34}.$$

## Epilogue or Paradigm Change

Sooner or later one comes to that
dreadful universal thing called
human nature.

**Oscar Wilde**

Paradigms change, though this misfortune happens very rarely in history. Man's psyche is the most resistant material across the Cosmos. The problem is that in sciences knowledge becomes belief. The complexity of the modern theoretical physics gave birth to a quite primitive belief about some monsterous mathematics as if inevitable in unified theories.

The Mayan priests invented a hieroglyphic system of writing unaccessible to ordinary human mind. However, this was the most profitable business for the intellectual elite of that society. The nation had no hope for simplifying their writing and reading practice. As a result, Mayan civilization stagnated and eventually collapsed. Complexity of the knowledge system makes civilization inept in critical moments.

The world is subtle. Complexity of the visible reality is the result of the simplicity of fundamental principles. God is by no means a fool who would risk to confuse Himself with whatever bizarre principles. And, this is called sophistication, that is, the wit to design complex systems out of nothing paricular such as spontaneity and bifurcation.

We have a scheme

**Aesthetics of Logic and Mathematics**

$$\Updownarrow$$

$$\{\textit{Simplicity of Foundations} \times \textit{Complexity of system}\} = \text{Sophistication}$$

Sophistication means that the cosmological reality can be even more complicated and mysterious than we imagine. Yet, the advantage of the superunification theory is notably that any mystery becomes decypherable by rational methods. According to an ancient myth, this year is 15804th in the divine chronology of human civilization:

(1111)
$$\sqrt[(3+1)]{\sqrt[\dim_\Phi \Phi^3 i_1]{\{\mathbf{15804} \times \mathbf{365.256}\}^{\pi \frac{\Phi^3 i_1}{\text{Spin}}}}} = \\ = \frac{10^{15}}{\mathbf{1 \cdot 2 \cdot 3 \cdot 4 \cdot 5 \cdot 6 \cdot 7 \cdot 8 \cdot 9}}.$$

If it is true that the cosmic time rhythm is defined by 7328, then

(1112)
$$\sqrt[\dim_\Phi \Phi^3 i_1 (3+1)]{\{\mathbf{15804} \cdot 365.256 \times \mathbf{7328}\}^{\pi \frac{\Phi^3 i_1}{\text{Spin}}}} = \mathbf{68631} \cdot 10^{10},$$

and

(1113)
$$\frac{\Phi^3 i_1 \cdot 10^9}{\mathbf{7328} \cdot 365.256} = 2013.$$

These considerations contradict to everything exempt Goedel's metatheorem and, therefore, they need neither in proof, not in refutation. They remain simply a fact of the cognitive history of civilization.

In 1998 at the beginning of this research I gave a talk to a group of scientists. I heard as a prominent theoretician wispered to the ears of his neighbour: The guy is going to play in numbers. I was quick to react and ask: What do you think God plays with?

Two-three years later the same theorist was keen to persuade people that everyone can do anything with numbers. And people do usually agree with such an ideology. Yet, playing in numbers is the most risky, practially hopeless, business. Either you find something eternal, or nothing. That's why theoreticians do never engaged in this business. God took this ultimate risk and succeeded. Any objection?